Maintenance Excellence

MECHANICAL ENGINEERING
A Series of Textbooks and Reference Books

Founding Editor

L. L. Faulkner

*Columbus Division, Battelle Memorial Institute
and Department of Mechanical Engineering
The Ohio State University
Columbus, Ohio*

Additional Volumes in Preparation

Mechanical Engineering Software

Maintenance Excellence

Optimizing Equipment Life-Cycle Decisions

edited by

John D. Campbell
PricewaterhouseCoopers LLP
Toronto, Ontario, Canada

Andrew K. S. Jardine
University of Toronto
Toronto, Ontario, Canada

MARCEL DEKKER, INC. NEW YORK · BASEL

ISBN: 0-8247-0497-5

This book is printed on acid-free paper.

Headquarters
Marcel Dekker, Inc.
270 Madison Avenue, New York, NY 10016
tel: 212-696-9000; fax: 212-685-4540

Eastern Hemisphere Distribution
Marcel Dekker AG
Hutgasse 4, Postfach 812, CH-4001 Basel, Switzerland
tel: 41-61-261-8482; fax: 41-61-261-8896

World Wide Web
http://www.dekker.com

The publisher offers discounts on this book when ordered in bulk quantities. For more information, write to Special Sales/Professional Marketing at the headquarters address above.

Preface

For the last 30 years, maintenance management has become increasingly sophisticated, at its highest levels becoming physical asset management. Physical asset management is seen as part of an interdependent system, a part that has a significant impact on profit, personnel, and the environment. As organizations seek to produce more, in less time, using fewer resources, physical asset management takes on a more central importance, requiring the management of a variety of factors to optimize asset life-cycle decisions. Second, sweeping changes in information technology have enabled a better, more complete interrogation of data, such that specialized software can efficiently transform that data into meaningful information through specific mathematical models.

In an environment of global competition, organizations

need to draw on specialized knowledge more than ever before. Specialists need to provide best practices to those organizations, and so need to be familiar with all aspects of physical asset management: why certain problems occur, what to do about those problems, and how to develop an organization's strategies to get the best results. With the advances in physical asset management discussed above, we have seen the opportunity for a new level of service to organizations that can make best use of the latest methodologies to optimize their operations, to find the best practices to get the best results. In seeking to provide that level of service, we have synergized our strengths into a constructive balance: one of us (Campbell) has extensive expertise in maintenance management consultation, with a focus on developing and applying systems that will best serve an organization; the other (Jardine) has a background as an innovator of the models needed to drive maintenance optimization. Together, and with some of the leading consultants at PricewaterhouseCoopers, we have been able to consolidate the talents and knowledge of a variety of individuals from around the world (including the contributors to this text) so that the service we deliver is at the leading edge of physical asset management concepts and methods. One result of that consolidation is the creation of this text.

The general aim of the book is to communicate an academic rigor with a practical direction. Applying physical asset management requires two key elements for the on-site manager: education and training. This volume is a tool to accomplish the former, providing a clear, thorough, and accessible expression of the theories and methodologies that are key to physical asset management. The appendices provide the mathematical explanations for those concepts, ensuring that the reader has the full picture, if required. With the background provided by the text, the manager may then have staff trained for the specific needs of his or her organization: ongoing trouble-shooting and the assessment of methodologies in pursuit of best practices. Alternatively, a manager could obtain external support to provide those services and to provide knowledge on the latest advances throughout the world.

The chapters provide focus to many key aspects of the field, as you will see. We trust that it will facilitate the task of any Physical Asset Manager in creating a path to maintenance excellence.

John D. Campbell
Andrew K. S. Jardine

Acknowledgments

We gratefully acknowledge the efforts of many dedicated colleagues in business and industry who contributed to the creation of this text.

Some of those colleagues work within the PricewaterhouseCoopers global family. Phil Upshall in London provided his wealth of experience in the large, complex worlds of utilities and public-sector organizations, including the postal services and defense. Rex Honey in Sydney, who has seen more mining operations around the world than I am sure he wishes to count, lent that expertise for our fleet and fixed-plant material. Michael Hawdon in Singapore has been applying reliability-based maintenance solutions to manufacturing and processing, integrating with just-in-time and total quality business environments. Finally, James Warner in the United

Kingdom and Mike Stoneham, Ralph Gardiner, and Tudor Negrea in Canada provided much-needed encouragement and support for the physical asset management practice.

The assistance of our business partners and maintenance management and reliability engineering professionals has been crucial to presenting an important balance in this book. *Plant Engineering and Maintenance Magazine*, by Clifford-Elliot Publishers, allowed us to use previously published materials by our authors and editors from their *Reliability Handbook*. We also received important contributions from individuals at some of the leading Enterprise Asset Management system vendors, notably Neil Cooper at Indus, Ted MacDonald at WonderWare, and Nick Beaton at Mincom.

In addition, we express our deepest appreciation to our clients for providing the battleground where these concepts, methods, and tools received their trial by fire before being presented to the business world at large. We offer special thanks to Mel Williams at Cardinal River Coal for his foresight in beta-testing EXAKT for CBM optimization, and Ram Thulasiram, a reliability engineer with Campbell's Soup Company Limited, for his successful experiences in applications of EXAKT.

Finally, I must thank my personal editor, Andrew Tausz, for ensuring that we have all been singing from the same song sheet and not mixed too many metaphors.

John D. Campbell

Organizing a text of this nature is an exercise in synthesizing ideas into a coherent piece of work, involving theory, practical application, and clarity of expression. Many individuals, in addition to those listed above, contributed to the development of this volume, to both the structure of its presentation and the academic research behind its theories. For some of the writing and editing, David Cope provided suggestions and organized material for a number of the chapters. Graham Oliver offered valuable insights about optimization criteria and in his marketing material for the RelCode, PERDEC, and AGE/CON software packages. As well, Greg Lemaich had the unenviable task of carefully reviewing the complete text, including the mathematical details of the appendices, and preparing the drawings in final form. He did a fine job, and any errors that remain should be seen as mine.

The development of ideas in this text as a whole relies in no small way on the academic work of the past as it is redefined in the present. Readers may recognize some of the models used here from *Maintenance, Replacement and Reliability*, one of my own texts from some years ago. Looking back at that time, I would like to acknowledge the wonderful inspiration of Professor D. J. White and the late Professor S. Vajda; the former introduced me to field of maintenance optimization and the latter, among other roles, was editor of the series of books in which my book appeared. More recently, my ideas in the optimization of maintenance decisions have been sharpened through annual meetings of the International Foundation for Research in Maintenance (IFRIM), founded in 1986 by W. M. J. Geraerds, Professor Emeritus of Eindhoven University of Technology in the Netherlands. Those meetings are attended by a fine group of academics from around the world whose teaching and research are focused on physical asset management.

Finally, I would like very much to acknowledge the value of almost daily discussions with colleagues in the CBM Laboratory at the University of Toronto, especially Professor V. Makis and Dr. D. Banjevic.

Andrew K. S. Jardine

Contents

Managing Equipment Reliability

Optimizing Maintenance Decisions

Conclusion

Contributors

Contributors

J. D. Campbell (Chapters 1 and 13)
A. K. S. Jardine (Chapters 10 and 11)
J. Kaderavek (Chapter 6)
M. Petit (Chapter 5)
J. V. Picknell (Chapters 2 and 7)
J. Shanahan (Chapter 4)
B. Stevens (Chapter 3)
D. Stretton (Chapter 8)
G. Walker (Chapter 6)
M. Wiseman (Chapters 9 and 12)

BIOGRAPHIES

J. D. Campbell   John Campbell is the Global Partner of PricewaterhouseCoopers' Physical Asset Management practice, based in Toronto. Mr. Campbell is a professional engineer. He graduated from the University of Toronto with a degree in Metallurgical Engineering and Materials Science, and is a Certified Management Consultant in operations management. Specializing in maintenance and materials management, he has over 20 years Canadian and international consulting experience in the assessment and implementation of strategy, management, and systems for maintenance, materials, and physical asset life-cycle functions. He authored the book *Uptime*, published in 1995, and is a coauthor of *Planning and Control of Maintenance Systems: Modeling and Analysis*, published in 1999.

A. K. S. Jardine   Andrew Jardine is a Professor in the Department of Mechanical and Industrial Engineering at the University of Toronto and principal investigator in the Department's Condition-Based Maintenance Laboratory, where the EXAKT software was developed. He also serves as a Senior Associate Consultant in the PricewaterhouseCoopers' global practice in Physical Asset Management. He has a Ph.D. from the University of Birmingham, England. Dr. Jardine is the author of the AGE/CON and PERDEC life-cycle costing software that is licensed to organizations including transportation, mining, electrical utilities, and process industries. He also wrote the book *Maintenance, Replacement and Reliability*, first published in 1973 and now in its sixth printing. In addition to being a sought-after speaker, he is a recognized authority in the world of reliability engineering and in the optimization of maintenance decision making. Dr. Jardine was the 1993 Eminent Speaker to the Maintenance Engineering Society of Australia and in 1998 was the first recipient of the Sergio Guy Memorial Award from the Plant Engineering and Maintenance Association of Canada, in recognition of his outstanding contribution to the maintenance profession.

J. Kaderavek   Joe Kaderavek is a Senior Consultant in the PricewaterhouseCoopers' global practice in Physical Asset Management, based in Sydney, Australia. He is a professional engineer and holds a B.Eng. degree from Sydney University, an M.B.A. degree from Deakin University, and postgraduate qualification in Reliability Engineering. He has over 10 years of experience, primarily in aircraft maintenance and logistics, but also including consulting in facility and plant maintenance management, strategy development and implementation, benchmarking for best practices, maintenance process redesign, life-cycle management, maintenance diagnostic assessments, and Computerized Maintenance Management Systems (CMMS) evaluation and implementation. He has worked with a range of industries including aerospace, defense, transport, mining, and consumer goods manufacturing.

M. Petit   Monique Petit is a Senior Consultant in PricewaterhouseCoopers' global practice in Physical Asset Management, based in Calgary. She has 20 years of experience in Materials and Production Management in various industries, and has spent the last 8 years on installation and process reengineering for numerous clients implementing Materials Modules of EAM (Enterprise Asset Management) Systems. She has a B.A. from the University of Lethbridge and has completed the Pulp and Paper Technician and Object-Oriented Client-Server Development certificate programs at NAIT. Ms. Petit has experience in process reengineering, physical warehouse layout, Material Management Operations optimization, and implementing inventory control best practices.

J. V. Picknell   James Picknell is a Director in the PricewaterhouseCoopers' global practice in Physical Asset Management, based in Toronto. He is a professional engineer and honors graduate of the University of Toronto with a degree in Mechanical Engineering, with further postgraduate studies at the Royal Naval Engineering College, the Technical University of Nova Scotia, and Dalhousie University. He has over 21 years of engineering and maintenance experience, including international consulting in plant and facility maintenance management, strategy development and implementation, reliability engineering, spares inventories, life-cycle costing and analysis, strategic diagnostic assessment, benchmarking for best practices, maintenance process redesign, and implementation of Computerized Maintenance Management Systems (CMMS). Mr. Picknell has worked in a range of industries, including defense, aerospace, marine, pulp and paper, water utility, automotive and consumer goods manufacturing, postal services, petrochemical, pharmaceutical, facilities management, mining, health care, and higher education.

J. Shanahan   John Shanahan is a Principal Consultant at PricewaterhouseCoopers working in the practice area of Physical Asset Management (PAM). He is a Chartered Electrical Engineer and a Member of the Institute of Electrical Engineers (U.K.) and the Project Management Institute, and holds an M.B.A. postgraduate degree. Mr. Shanahan is responsible for all Enterprise Asset Management (EAM) systems aspects of the PAM practice from strategic planning to postimplementation audits. Typical services packages offered by the practice include business case development, functional requirements definition, system selection, process mapping and optimization, preimplementation readiness assessments, complete implementation services, postimplementation audits, and benefits realization. Mr. Shanahan has an extensive background in large project proposal development and project management, deriving from a career in Electrical Engineering Systems and Information Technology projects. He has managed projects in many industry sectors including heavy construction, utilities, military logistics and maintenance, and Information Technology.

B. Stevens   Ben Stevens is a Managing Associate in PricewaterhouseCoopers' global practice in Physical Asset Management. He was educated at the University of Sheffield, England, is an honors graduate in Economics and Accounting, and holds a Master's degree in Managerial Economics. He has over 30 years of experience, with the past 12 dedicated to the marketing, sales, development, selection, justification, and implementation of Computerized Maintenance Management Systems. His prior experience includes the development, manufacture, and implementation of production monitoring systems, executive level management of maintenance, finance, administration functions, management of reengineering efforts for a major Canadian bank, and management consulting. He has worked in and with a range of industries, including CMMS, defense, petrochemical, health care, waste management, pulp and paper, high tech electronics, nuclear, textile manufacturing, and finance. He is a frequent speaker at conferences and training sessions, and he chaired the International CMMS conferences in Dubai in 1996, 1997, 1998, and 1999.

D. Stretton   Doug Stretton is a Senior Consultant in PricewaterhouseCoopers' global practice in Physical Asset Management, based in Toronto. He is a Professional Engineer and an honors graduate of Queen's University, Canada, in Mechanical Engineering. After gaining valuable work experience he returned to Queen's University and earned an M.B.A. He has 10 years of experience in progressive positions in Maintenance, Engineering, and Manufacturing Engineering. Mr. Stretton has specialized in turnarounds of maintenance departments and production lines using Total Productive Maintenance (TPM) and continuous improvement (CI) techniques. He has worked in a variety of industries including steel, food, consumer household products, printing, and nonferrous metals.

G. Walker   Geoff Walker is leader of PricewaterhouseCoopers' Physical Asset Management practice in Europe, the Middle East, and Africa, which integrates maintenance management with spares management and procurement to provide integrated solutions for asset intensive industries, particularly those with widely distributed networks of assets such as the utilities and defense sectors. He has helped a wide range of organizations to improve performance and reduce costs through adopting best practice across the entire engineering supply chain. Before joining Coopers & Lybrand, Geoff Walker spent 12 years in Engineering and Production Management in the chemicals industry, which included Maintenance, New Projects, and Technology Development in a range of chemical plants. His experience covers the entire range of chemical production from initial laboratory synthesis through pilot plant and small scale manufacturing to full-scale chemical synthesis and final formulation, filling, and packing into sales containers. He is a chartered mechanical engineer, and holds an M.A. degree in Engineering from the University of Cambridge and an M.B.A. from the University of London.

M. Wiseman Murray Wiseman is a Principal Consultant in PricewaterhouseCoopers' global practice in Physical Asset Management, based in Toronto. He has been in the maintenance field for more than 18 years. His previous responsibilities and roles include maintenance engineering, planning, and supervision, as well as founding and operating a commercial oil analysis laboratory. He developed a Web-enabled Failure Modes and Effects Criticality Analysis (FMECA) system conforming to MIL STD 1629A, in which he incorporated an expert system as well as links to two failure rate/failure mode distribution databases at the Reliability Analysis Center in Rome, New York. Mr. Wiseman's particular strength and interest are in the CBM optimization area, applying modeling techniques when the RCM process signals that condition monitoring is the appropriate tactic. Mr. Wiseman holds a B. Eng. Mech. degree from McGill University, Montréal, and is a member of the Pulp and Paper Institute of Canada.

Maintenance
Excellence

1

Introduction

Maintenance excellence is many things, done well. It's when a plant performs up to its design standards and equipment operates smoothly when needed. It's maintenance costs tracking on budget, with reasonable capital investment. It's high service levels and fast inventory turnover. It's motivated, competent trades.

Most of all, maintenance excellence is the balance of performance, risk, and cost to achieve an optimal solution. This is complicated because much of what happens in an industrial environment is by chance. Our goal is exceptional performance. That isn't made any easier by the random nature of what we're often dealing with.

Maintenance management, though, has evolved tremendously over the past century. As Coetzee (1) has said, it wasn't even contemplated by early equipment designers. Since then, Parkes (2) has described how maintenance evolved initially from uncomplicated and robust equipment to built-in obsolescence. From Kelly (3) we learned the progression from preventive and planned maintenance after World War II to industrial engineering, industrial psychology, reliability, and operations research in the '60s and '70s to condition monitoring, computerization, and life-cycle management in the '80s. Moubray (4) described how our expectations have changed with the generations, and how safety, environmental, and product quality issues are now as important as reliability. Campbell (5) has shown us the way in which equipment changes have advanced maintenance practices, with predominant tactics changing from run-to-failure to prevention, then prediction and eventually reliance. It's the reliance tactic that is the focus of this book.

Many organizations developed a systematic approach to maintenance planning and control in the late '70s and early '80s, only to abandon it as we shifted to a global marketplace and major economic harmonization. Today we must start over. We must first rebuild basic capabilities in maintenance management, before we can prove the value of reliability management and maintenance optimization.

There are three types of goals on the route to maintenance excellence:

- **Strategic**. First, you must draw a map and set a course for your destination. You need a vision of what maintenance management will be like in, say, three years. What is the plant condition, the availability, the maintenance cost structure, the amount of planned work compared to unplanned reactive work, the work environment? You must assess where you are today, to get where you are going. This way, you will know the size of the gap to be closed. Finally, you must determine the human, financial, and physical resource requirements, as well as a timeframe, to make your vision real.

- **Tactical**. Now, you need a work management and materials management system to control the maintenance process. Ideally, this is a computerized maintenance management system, an enterprise asset management system, or a maintenance module in an enterprise resource planning system. Maintenance planning and scheduling—for job work orders, plant and equipment shutdowns, and annual budgeting exercises, and for creating a preventive and predictive program—is most important. Also, measure performance at all levels, to effectively change people's behavior and to implement lubrication, inspection, condition-monitoring, and failure-prevention activities.

- **Continuous improvement**. Finally, if you engage the collective wisdom and experience of your entire workforce, and adapt "best practices" from within and outside your organization, you will complete the journey to systematic maintenance management. But continuous improvement requires diligence and consistency. To make it work, you need a method, a champion, strong management, and hard work.

MAINTENANCE EXCELLENCE FRAMEWORK

As you read on, you will learn how to manage equipment reliability and to optimize maintenance—the life cycle of your plant, fleet, facility, and equipment. The purpose of this book is to provide a framework. It is divided into four sections: Maintenance Management Fundamentals; Managing Equipment Reliability; Optimizing Maintenance Decisions, and Achieving Maintenance Excellence.

Maintenance Excellence

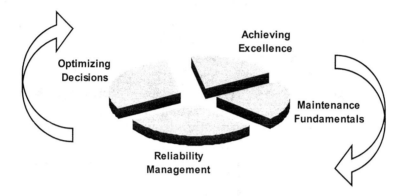

Figure 1-1   Components of maintenance excellence.

PART I: MAINTENANCE MANAGEMENT FUNDAMENTALS

Chapter 2 provides an overview of the basic strategies, processes, and approaches for managing equipment reliability
through work management and leading maintenance management methods and tools.

Chapter 3 explores the concept that if you can't measure
it, you can't manage it. You'll learn how to monitor and control
the maintenance process. We discuss the top-down approach,
from setting strategic business plans to ensuring that maintenance fully supports them. We also look at the measures
needed on the shop floor to manage productivity, equipment
performance, cost management, and stores materials management. When you pull it all together, it becomes a balanced
scorecard.

Chapter 4 discusses the latest computerized maintenance
management and enterprise asset management systems. We
describe the basics of determining requirements, then justifying, selecting, and implementing solutions that realize the

benefits. Materials are the single biggest expense for most maintenance operations, so it's no wonder that poor maintainability is usually blamed on a shortcoming in parts, components, or supply.

Chapter 5 describes how you can manage maintenance procurement and stores inventory to support effective and efficient work management and equipment reliability. We show you how to invest the limited parts inventory budget wisely to yield both top service levels and high turnover.

PART II: MANAGING EQUIPMENT RELIABILITY

Chapter 6 is about assessing risks and managing to international standards. We begin by looking at what equipment is critical based on several criteria, then develop methods for managing risk. To help you develop guidelines and easily assess the state of risk in an enterprise, we describe relevant international standards for managing risk.

Chapter 7 summarizes the Reliability-Centered Maintenance methodology as the most powerful tool to manage equipment reliability. This often misunderstood yet incredibly powerful approach improves reliability while helping to control maintenance costs in a sustainable way.

Chapter 8 examines reliability by the operator—what many leading companies are calling their total productive maintenance programs. We describe how equipment management and performance is everyone's job, especially the operator's.

PART III: OPTIMIZING MAINTENANCE DECISIONS

Chapter 9 provides the basics for using statistics and asset life cost in maintenance decision making. We hear about a company's 6 sigma quality management program, but the link between sigma, or standard deviation, and maintenance management is often subtle, at best. In this chapter, we show that collecting simple failure frequency can reveal the likelihood of when the next failure will happen. While exploring failure probabilities, we guide you through the concept of life-cycle costs and discounting for replacement equipment investments.

With the fundamentals in place and ongoing effective reliability management, Chapter 10 begins the journey to optimization. Here we explore the use of mathematical models and simulation to maximize performance and minimize costs over equipment life. You'll find an overview of the theory behind component replacement, capital equipment replacement, inspection procedures, and resource requirements. Various algorithmic and expert system tools are reviewed along with their data requirements.

Chapter 11 focuses further on critical component and capital replacements. We include engineering and economic information in our discussion about preventive age-based and condition-based replacements.

Chapter 12 takes an in-depth look at how condition-based monitoring can be cost-effectively optimized. One of the biggest challenges facing the Maintenance Manager is figuring out what machine condition data to collect, and how to use it. We describe a statistical technique to help make practical decisions for run-to-failure, repair, or replacement, including cost considerations.

PART IV: ACHIEVING MAINTENANCE EXCELLENCE

In the last chapter, Chapter 13, we show how to apply the concepts and methods in this book to the shop floor. This involves a three-step process. The first is to determine your current state of affairs, the best practices available, and your vision. You need to know the size of the gap between where you

are and where you want to be. The second step involves building a conceptual framework and planning the concepts and tools to execute it. Finally, we look at the implementation process itself, and all that goes into managing change.

THE SIZE OF THE PRIZE

We could debate at length about the social and business mandate of an organization with a large investment in physical assets. But this is sure: if these assets are unproductive, it becomes irrelevant. Productivity is what you get out for what you put in. Maintenance excellence is about getting exemplary performance at a reasonable cost. What should we expect for investing in maintenance excellence? What is the size of the prize?

Let's look at capacity. One way to measure it is:

$$\text{Capacity} = \text{Availability} \times \text{Utilization} \times \text{Process rate} \times \text{Quality rate}$$

If the equipment is available, being used, running at the desired speed, and precise enough to produce the desired quality and yield, we have the required maintenance output.

Now look at cost. This is a bit more difficult, because cost can vary depending on many things: the working environment, equipment age and use, operating and maintenance standards, technology, etc. One thing we know: a breakdown maintenance strategy is more costly than linking maintenance actions to the likely failure causes.

But by how much? Here's a helpful rule of thumb to roughly estimate cost-saving potential in an industrial environment:

$$\text{\$1 Predictive, Preventive, Planned} = \text{\$1.5 Unplanned, Unscheduled} = \text{\$3 Breakdown}$$

Accomplishing "one unit" of maintenance effectiveness will cost $1 in a planned fashion, $1.50 in an unplanned way, and $3 if reacting to a breakdown. In other words, you can pay now or you can pay more later. Emergency and breakdown maintenance is more costly for a number of reasons:

- You must interrupt production or service.
- The site isn't prepared.
- Whoever is available with adequate skills is pulled from his or her current work.
- You must obtain contractors, equipment rental, and special tools quickly.
- You have to hunt down or air-freight in materials.
- The job is worked on until completed, often with overtime.
- There usually isn't a clear plan or drawings.

For example, if the total annual maintenance budget is $100 million and the work distribution is 50% planned, 30% unplanned and unscheduled, and 20% breakdown:

$$(50\% \times 1) + (30\% \times 1.5) + (20\% \times 3) = 50 + 45 + 60$$
$$= 155 \text{ "equivalent planned units"}$$

Planned work costs $\dfrac{50}{155} \times \$100$ million

$$= \$32 \text{ million, or } \$.645 \text{ million per unit}$$

Unplanned work costs $\dfrac{45}{155} \times \$100$ million $= \$29$ million

Emergency work costs $\dfrac{60}{155} \times \$100$ million $= \$39$ million

To compare the difference, imagine maintenance improvement yielding 60% planned, 25% unplanned and 15% breakdown:

Planned work costs $.645 \times 1 \times 60\% = \$39$ million

Unplanned work costs $.645 \times 1.5 \times 25\% = \$24$ million

Emergency work costs $.645 \times 3.0 \times 15\% = \$29$ million

Total $= \$92$ million

Savings potential $= \$100$ million $-\$92$ million $= \$8$ million

The challenge of maintenance excellence, and the goal of this book, is to develop tactics that will minimize breakdowns and maximize the rewards of planned, preventive, and predictive work.

REFERENCES

1. Jasper L. Coetzee. Maintenance. South Africa: Maintenance Publishers, 1997.
2. D. Parkes. Maintenance: Can it be an exact science? In: A. K. S. Jardine, ed. Operational Research in Maintenance. Manchester University Press/Barnes and Noble, Inc., 1970.
3. A. Kelly. Maintenance and Its Management. Farnham, England: Conference Communication, 1989.
4. J.M. Moubray. Reliability-Centered Maintenance. 2d ed. Oxford, England: Butterworth Heinemann, 1997.
5. John Dixon Campbell. Uptime: Strategies for Excellence in Maintenance Management. Portland, OR: Productivity Press, 1995.

2

Maintenance Management Methodologies

Today's maintenance and physical asset managers face great challenges to increase output, reduce equipment downtime, and lower costs, and do it all with less risk to safety and the environment. This chapter addresses the various ways to accomplish these objectives by managing maintenance effectively, within your organization's unique business environment.

Of course, you must make tradeoffs, such as cost versus reliability, to stay profitable in current markets. We show how to balance the demands of quality, service, output, costs, time, and risk reduction. This chapter also examines just how maintenance and reliability management can increase profits and add real value to the enterprise.

We discuss the levels of competence you must achieve on the road to excellence. There are clear evolutionary development stages. To get to the highest levels of expertise, you must ensure that the basics are in place. How can you tell if you are ready to advance? The charts that follow will help you decide.

The final sections of the chapter describe the methods used by companies that truly strive for excellence—Reliability-Centered Maintenance, Root Cause Failure Analysis, and Optimization of Decision Making. This sets the stage for material presented later in the book.

2.1 MAINTENANCE: THE CHALLENGE

Smart organizations know they can no longer afford to see maintenance as just an expense. Used wisely, it provides essential support to sustain productivity and fuel growth, while driving down unneeded and unforeseen overall expenses. Effective maintenance aims to:

- Maximize uptime (productive capacity)
- Maximize accuracy (the ability to produce to specified tolerances or quality levels)
- Minimize costs per unit produced (the lowest cost practical)
- Minimize the risk that productive capacity, quality, or economic production will be lost for unacceptable periods of time
- Prevent safety hazards to employees and the public, as much as possible
- Ensure the lowest possible risk of harming the environment

In fact, in today's competitive environment, all of these are strategic necessities to remain in business. The challenge is how best to meet the above demands. In many companies, you have to start at the beginning—put the basics in place—before your attempts to achieve excellence and optimize decisions will be successful. The ultimate aim is to attain a high degree of control over your maintenance decisions, and this chapter explores various maintenance management methodologies to help you get there.

2.2 METHODOLOGIES

Optimization is a process that seeks the best solution, given competing priorities. This entails setting priorities and making compromises for what is most important. Maximizing profits depends on keeping our manufacturing assets in working order, yet maintenance sometimes requires downtime, taking away from production capacity. Minimizing downtime is essential to maximize the availability of our plant for production. Optimization will help you find the right balance.

Even though increasing profit, revenues, availability, and reliability while decreasing downtime and cost is all related, you can't always achieve them together. For example, maximizing revenues can mean producing higher-grade products that command higher prices. But that may require lower production volumes and therefore higher costs per unit produced.

Clearly, cost, speed, and quality objectives can compete with one another. An example is the improved repair quality from taking additional downtime to do a critical machine alignment correctly. The result will probably be longer run time before the next failure, but it does cost additional repair downtime in the short term.

The typical tradeoff choices in maintenance arise from trying to provide the maximum value to our "customers". We want to maximize:

- Quality (e.g., repair quality, doing it right the first time, precision techniques)
- Service level (e.g., resolution and prevention of failures)
- Output (e.g., reliability and uptime)

At the same time, we want to minimize:

- Time (e.g., response and resolution time and Mean Time to Repair (MTTR))
- Costs (e.g., cost per unit output)
- Risk (e.g., predictability of unavoidable failures)

Management methods seek to balance these factors to deliver the best possible value. Sometimes, however, you must educate the customer about the tradeoff choices you face, to ensure "buy-in" to the solution. For example, a production shift supervisor might not see why you need additional downtime to finish a repair properly. You have to convince him or her of the benefit—extended time before the next failure and downtime.

Maintenance and reliability are focused on sustaining the manufacturing or processing assets' productive capacity. By sustaining we mean maximizing the ability to produce quality output at demonstrated levels. This may mean production levels that are beyond original design if they are realistically sustainable.

2.3 WHERE DOES MAINTENANCE AND RELIABILITY MANAGEMENT FIT IN TODAY'S BUSINESS?

The production assets are merely one part of an entire product supply chain that produces profit for the company. It is important to recognize that maintenance priorities may not be most important for the company as a whole.

In a very basic manufacturing supply chain, materials flow from source (suppliers) through primary, and sometimes secondary, processing or manufacturing, then outbound to customers through one or more distribution channels. The traditional business focus at this level is on purchasing, materials requirements planning, inventory management, and just-in-time supply concepts. The objective is to minimize work in process and inventory while manufacturing, to ship for specific orders (the pull concept).

To optimize the supply chain, you optimize the flow of information backward, from customers to suppliers, to produce the most output with the least work in process. Supply chain optimizing strategies are to:

- Improve profitability by reducing costs
- Improve sales through superior service and tight integration with customer needs
- Improve customer image through quality delivery and products
- Improve competitive position by rapidly introducing and bringing to market new products

Methods to achieve these include:

- Strategic material sourcing
- Just-in-time inbound logistics and raw materials management
- Just-in-time manufacturing management
- Just-in-time outbound logistics and distribution management
- Physical infrastructure choices
- Eliminating waste to increase productive capacity
- Using contractors or outsource partners
- Inventory management practices

Business processes that are involved include:

- Marketing
- Purchasing
- Logistics
- Manufacturing
- Maintenance
- Sales

- Distribution
- Invoicing and collecting

At the plant level, you can improve the manufacture part of the process by streamlining production processes through just-in-time materials flows. This way, you'll eliminate wasted efforts and reduce the production materials and labor needed.

In the past, maintenance received little recognition for its contribution to sustaining production capacity. It tended to be viewed only as a necessary and unavoidable cost. Even at the department level today, managers typically don't view the big picture—the entire plant—they focus only on their departmental issues. Unfortunately, maintenance is often viewed only within the context of keeping down costs. In accounting, maintenance shows up as an operating expense, one that should be minimized. Disregarding the cost of raw materials, maintenance is typically only a fraction of manufacturing costs (5–40%, depending on the industry). Similarly, those manufacturing costs are a fraction of the products' total selling price.

Reducing maintenance expenses does indeed add to the bottom line directly but, since it is a fraction of a fraction of the total costs, it is typically seen as less important, commanding less management attention. Most budget administrators don't seem to fully understand maintenance, judging it by historical cost numbers. When you reduce a maintenance budget, service ultimately declines. Also, output is usually reduced and risk increased when there isn't enough time or money to do the work right the first time.

Of course, the accounting view is one-dimensional because it looks only at costs. When you consider the value that maintenance delivers, it becomes much more important. By sustaining quality production capacity and increasing reliability, you generate more revenue and reduce disruptions.

This requires the right application of maintenance and reliability. Of course, doing maintenance properly means being proactive and accepting some amount of downtime. Effective maintenance methods are needed to make the best possible use of downtime and the information you collect to deliver the best value to your production customers.

2.4 WHAT MAINTENANCE PROVIDES TO THE BUSINESS

Maintenance enhances production capacity and reduces future capital outlay. It does this by:

- Maximizing uptime
- Maximizing accuracy, producing to specified tolerances or quality levels
- Minimizing costs per unit produced
- Sustaining the lowest practical and affordable risk to loss of production capacity and quality
- Reducing as much as possible the safety risk to employees and the public
- Ensuring the lowest possible risk of harming the environment

Notice the emphasis on risk reduction. This is why insurers and classification societies take a keen interest in their clients' maintenance efforts. Your maintenance reduces their exposure to risk and helps keep them profitable. Nearly every time a major accident involves a train, airplane, or ship, there is an in-depth investigation to determine whether improper maintenance was the cause of the disaster.

Maintenance can also provide a strategic advantage. Increasingly, as companies automate production processes and manufactured goods are treated like commodities, the lowest-cost producer will benefit. Automation has reduced the size of production crews while increasing the amount and complexity of work for maintenance crews. Maintenance costs will therefore increase relative to direct production costs. Even low-cost producers can expect maintenance costs to rise. That increase must be offset by increased production. You need less downtime and higher production rates, as well as better quality at low unit cost—that means more effective and efficient maintenance.

Achieving all this requires a concerted effort to *manage* and *control* maintenance rather than letting the assets and their random failures control costs. In today's highly competitive business environment, you cannot afford to let that happen. Unfortunately, many companies do just that, allowing natural processes to dictate their actions. By operating in a "firefighting" mode, they merely respond to rolls of the dice, with random results. Without intervening proactively, these companies can only react after the fact, once failures occur. The consequences are low reliability, availability, and productivity—ingredients for low profitability.

2.5   READY FOR EXCELLENCE?

Optimizing your effectiveness cannot be accomplished in a chaotic and uncontrolled environment. Optimization entails making intelligent and informed decisions. That involves gathering accurate and relevant information to support decisions and acting in a timely manner. As the saying goes, "When up to the rear in alligators, it's difficult to remember to drain the swamp." You must have your maintenance system and process under control before you can optimize effectively. You need to tame the alligators with good maintenance management methods, followed in a logical sequence.

Campbell (1) teaches that several elements are necessary for maintenance excellence to be achieved. They fall into three major areas:

- Exercise leadership at all times. Without it, change won't be successful.

- Achieve control over the day-to-day maintenance operation.
- Apply continuous improvement, once you have control, to remain at the leading edge of your industry.

Leadership includes:

- Organization structure and style. It must fit the business environment and the strategy you're implementing. It can be centralized along shop lines, fully distributed along area lines, or some blend of the two. It must be consistent with business objectives. If effective teamwork is called for, the structure must reflect the methods to achieve it. Of course, the number of maintainers you use must match the workload. If your organization is understaffed, you'll be stuck in a reactive firefighting mode. If you overstaff, it will be too expensive.
- A clear future vision for the organization, with a high-level plan to achieve it. Know your objectives before making changes. Make sure that the entire maintenance department, along with operations and engineering, understands them too. You need good communication so that employees buy into the vision. If they don't, implementing it successfully will be nearly impossible. Strategy is what gets done, not what is said.
- A method to manage the transition to the future. Change is always difficult to achieve successfully. We all get used to the way things are, even if they could be better. You have to be able to motivate your people. They need to understand why it is important to change and what the impact will be. They need to see management's commitment to making it happen, as well as measurements that show progress. Finally, they need to know what's in it for them personally.

To paraphrase Kotter (2), leaders make change happen and managers keep things running smoothly. To improve and optimize maintenance management, you need a strong leader to champion change.

Control entails:

- Planning and scheduling practices to manage service delivery. Through careful planning, you establish what will be done, using which resources, and provide support for every job performed by the maintainers. You also ensure that resources are available when needed. Through scheduling you can effectively time jobs to decrease downtime and improve resource utilization.
- Materials management practices to support service delivery. Part of the job of a planner is to ensure that any needed parts and materials are available before work starts. You can't make a schedule until you know for certain the materials are available. To minimize operations disruptions, you need spare parts and maintenance materials at hand. Effective materials management has the right parts available, in the right quantities, at the right time, and distributes it cost-effectively to the job sites.
- Maintenance tactics for all scenarios: to predict failures that can be predicted, prevent failures that can be prevented, run-to-failure when safe and economical to do so, and to recognize the differences. This is where highly technical practices such as vibration analysis, thermographics, oil analysis, nondestructive testing, motor current signature analysis, and judicious use of overhauls and shutdowns are deployed. These tactics increase the amount of preventive maintenance that can be planned and scheduled and reduce the reactive work needed to clean up failures.

- Measurements of maintenance inputs, processes, and outputs to help determine what is and isn't working, and where changes are needed. We all behave according to the way we are measured. We adjust our driving speed and direction to signs and gauges. We manage our investments to minimize income taxes. In maintenance, we deploy more preventive and predictive methods to reduce the volume of reactive work. By measuring your inputs (costs) and outputs (reliability or uptime), you can see whether your management is producing desired results. If you also measure the processes themselves, you can control them more closely and adjust them to increase your success.
- Systems that help manage the flow of control and feedback information through these processes. Accounting uses computers to keep their books. Purchasing uses computers to track their orders and receipts and to control who gets paid for goods received. Likewise, Maintenance needs effective systems to deploy the workforce on the many jobs that vary from day to day, and to collect feedback to improve management and results.

Continuous-Improvement RCM and TPM methods are described in detail in Chapters 7 and 8.

The degree to which a company achieves these objectives indicates its level of maturity. A maturity profile is a matrix that describes the organization's characteristic performance in each of these elements. One example appears as Figure 1-5 in Campbell's book (1). Figure 2-1 here presents another example of a profile that covers the spectrum of elements needed for maintenance excellence. It is presented in a series of profiles, with supporting details for a cause-and-effect diagram. Every leg of the diagram comprises several elements, each of which can grow through various levels of excellence:

Elements of Excellence

Figure 2-1 Elements of maintenance excellence.

- Novice
- Foundation
- Apprentice
- Journeyman
- Mastery

Figures 2-2 to 2-7 describe every level for each element from Figure 2-1. The color (gray scale) intensity expresses degrees of excellence, with the deeper shades representing greater depth and breadth. Leadership and people are the most important elements, although they are not always treated as such.

As you can see, the organization moves from reactive to proactive, depends more heavily on its employees, and shifts from a directed to a more autonomous and trusted workforce. Organizations that make these changes often use fewer people to get as much, or more, work done. They are typically very productive.

Leadership and People

Leadership & People
Strategy & Business Planning →
Organization & Numbers →
Training, Skills, Knowledge & Ability →
Motivation & Change Readiness →
Autonomy, Teamwork →

Maintenance Excellence

Leadership & People	Strategy & Business Planning	Organization & Numbers	Training, Skills, Knowledge & Ability	Motivation & Change Readiness	Autonomy & Teamwork
Mastery	Stated strategy with mission, long range vision, goals. Goals are specific, measurable, achievable, realistic and timed (for 2 or more years). Actions match words. Strategy linked with corporate goals.	Decentralized teams operate independent of daily maintenance control & may report to production. Plenty of interaction with production crew members. Maintenance supports teams.	Trades are largely multi-skilled with some multi-trade qualified individuals and regularly use their qualifications. Production staff do minor equipment upkeep tasks. Training time at least 2 weeks per trade per year.	Trades' compensation has a reward component linked to business results. Competitive forces widely accepted as driving need for beneficial changes. Changes initiated by both management and workforce. Changes are usually successful and measurable benefits achieved.	Decentralized teams are self directed and base decisions on business need. Excellent cooperation between maintenance and production at all levels. Teamwork is a visible hallmark of the entire organization.
Journeyman	Strategy (as above) but not linked to corporate goals. Actions close to the words.	Decentralized teams controlled by maintenance have plenty of interaction with production crew members.	Trades are largely multi-skilled and regularly use their skills. Production staff do some minor equipment upkeep tasks. Training time 1 to 2 weeks per trade per year.	Cooperative atmosphere prevails, trust between management and labour is high. Change always initiated by management and the need for changes explained in advance and widely accepted. Changes are usually successful.	Some self directed workers and teams. Good cooperation between production and maintenance at all levels. Teamwork may be a feature of the entire organization.
Apprentice	Some goal setting for long term, annual plans used.	Mix of decentralized teams reporting to maintenance and central shop structure.	Trades have some multi-skilling and often use those skills. Production staff do minimal minor upkeep tasks. Training time less than 1 week per trade per year. Training need analysis completed for all trades.	Some cooperation between management and labour exists and level of trust is moderate. Reason for change is usually explained in advance. Changes sometimes fail.	Directed workforce with some teamwork but little to no team training. Some cooperation between maintenance and production at the working level.
Foundation	PM program in place, benefits recognized.	Centralized structure based on trades breakdown. Control through maintenance supervisors / leads in response to production demands.	No multi-skilling is used. Production staff do no equipment upkeep. Training time less than 1 week per trade per year. Some training need analysis performed.	Management motivation explained when questioned. Some distrust but desire to improve exists. Changes often fail.	Directed workforce with no attempt at teamwork outside of shop structure. Good cooperation between production and maintenance leadership.
Novice	Breakdown maintenance, fire fighting, no stated goals.	Centralized structure based on trades breakdown. Action directed largely by operations supervisors.	No multi-skilling is used. Production staff do no equipment upkeep. Training is driven by necessity only.	Highly resistive to change. Hourly workforce generally distrusts management motives. No visible desire to improve. Change initiatives usually fail.	Directed workforce with no attempt at teamwork outside of shop structure. Maintenance and production relationship is strained.

Figure 2-2 Leadership and people elements (who does it all).

The next most important elements are usually the methods and processes. These are all about how you manage maintenance. They are the activities that people in the organization actually do. Methods and processes add structure to the work that gets done. As processes become more effective, people become more productive. Poor methods and processes produce much of the wasted effort typical of low-performing maintenance organizations. Because of the extent of coverage, expertise, and capability needed, these levels of excellence are shown in two figures (2-3 and 2-4).

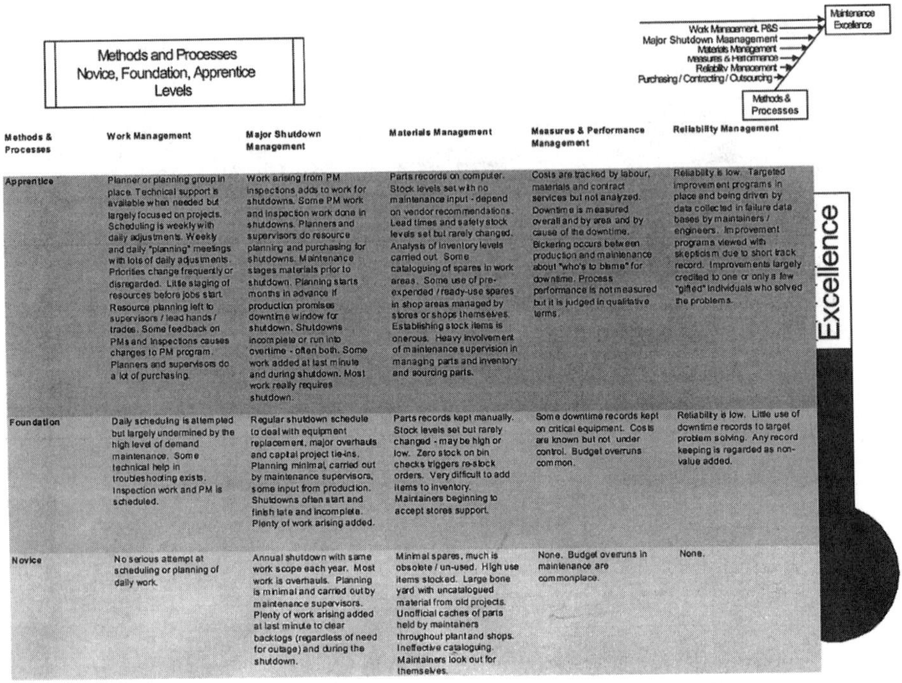

Methods & Processes	Work Management	Major Shutdown Management	Materials Management	Measures & Performance Management	Reliability Management
Apprentice	Planner or planning group in place. Technical support is available when needed but largely focused on projects. Scheduling is weekly with daily adjustments. Weekly and daily "planning" meetings with lots of daily adjustments. Priorities change frequently or disregarded. Little staging of resources before jobs start. Resource planning left to supervisors / lead hands / trades. Some feedback on PMs and inspections causes changes to PM program. Planners and supervisors do a lot of purchasing.	Work arising from PM inspections adds to work for shutdowns. Some PM work and inspection work done in shutdowns. Planners and supervisors do resource planning and purchasing for shutdowns. Maintenance stages materials prior to shutdown. Planning starts months in advance if production promises downtime window for shutdown. Shutdowns incomplete or run into overtime - often both. Some work added at last minute and during shutdown. Most work really requires shutdown.	Parts records on computer. Stock levels set with no maintenance input - depend on vendor recommendations. Lead times and safety stock levels set but rarely changed. Analysis of inventory levels carried out. Some cataloguing of spares in work areas. Some use of pre-expended / ready-use spares in shop areas managed by stores or shops themselves. Establishing stock items is onerous. Heavy involvement of maintenance supervision in managing parts and inventory and sourcing parts.	Costs are tracked by labour, materials and contract services but not analyzed. Downtime is measured overall and by area and by cause of the downtime. Bickering occurs between production and maintenance about "who's to blame" for downtime. Process performance is not measured but it is judged in qualitative terms.	Reliability is low. Targeted improvement programs in place and being driven by data collected in failure data bases by maintainers / engineers. Improvement programs viewed with skepticism due to short track record. Improvements largely credited to one or only a few "gifted" individuals who solved the problems.
Foundation	Daily scheduling is attempted but largely undermined by the high level of demand maintenance. Some technical help in troubleshooting exists. Inspection work and PM is scheduled.	Regular shutdown schedule to deal with equipment replacement, major overhauls and capital project tie-ins. Planning minimal, carried out by maintenance supervisors, some input from production. Shutdowns often start and finish late and incomplete. Plenty of work arising added.	Parts records kept manually. Stock levels set but rarely changed - may be high or low. Zero stock on bin checks triggers re-stock orders. Very difficult to add items to inventory. Maintainers beginning to accept stores support.	Some downtime records kept on critical equipment. Costs are known but not under control. Budget overruns common.	Reliability is low. Little use of downtime records to target problem solving. Any record keeping is regarded as non-value added.
Novice	No serious attempt at scheduling or planning of daily work.	Annual shutdown with same work scope each year. Most work is overhauls. Planning is minimal and carried out by maintenance supervisors. Plenty of work arising added at last minute to clear backlogs (regardless of need for outage) and during the shutdown.	Minimal spares, much is obsolete / un-used. High use items stocked. Large bone yard with uncatalogued material from old projects. Unofficial caches of parts held by maintainers throughout plant and shops. Ineffective cataloguing. Maintainers look out for themselves.	None. Budget overruns in maintenance are commonplace.	None.

Figure 2-3 Methods and processes (what gets done): the basic levels.

In Figure 2-5, systems and technology represent the tools used by the people implementing the processes and methods you choose. These are the enablers, and they get most of the attention in maintenance management. Some organizations that focus tremendous energy on people and processes, with only basic tools and rudimentary technology, still achieve high performance levels. Other organizations, focused on the tools shown in Figure 2-5, haven't. Generally, emphasizing technology without excellence in managing methods, processes, and people will bring only limited success. It's like the joke about needing a computer to really mess things up. If inefficient or ineffective processes are automated, and then run with inef-

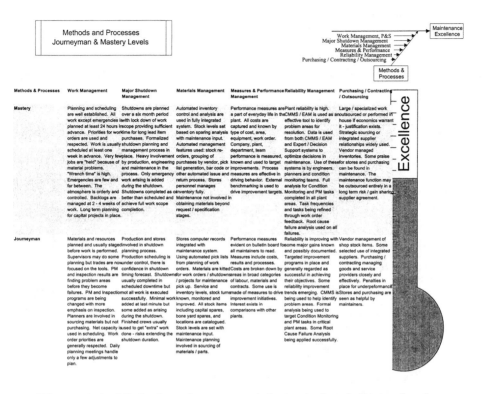

Figure 2-4 Methods and processes (what gets done): the top levels.

fective and unmotivated employees, the result will be disappointing.

Ultimately, how well you manage processes, methods, and the people who use them comes down to the materials and physical plant you maintain. Figures 2-6 and 2-7 depict levels of excellence. It is in the physical plant and materials area where you can best judge maintenance management effectiveness, which is why these descriptions are more detailed.

You can see that in these basic levels, plant and materials support to maintenance are in rough shape. Because capital management operates reactively, things are left to run down.

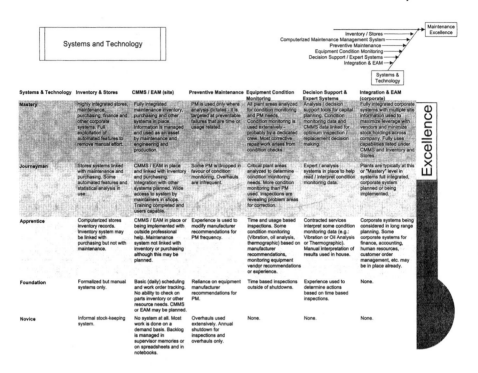

Figure 2-5 Systems and technology (the enablers).

In Figure 2-7, you can see that maintenance management is moving in a proactive direction and the plant is being sustained far better.

Optimization technologies and methods appear at the "journeyman" and "mastery" levels of excellence. It is unlikely that optimization would be successful at other levels, at least overall. You may need to undergo fairly broad change to gain control and introduce continuous improvement, as described in Ref. 1.

Just as you must learn to walk before you run, you must be in firm control of maintenance before you can successfully begin continuous improvement. You do this by incrementally

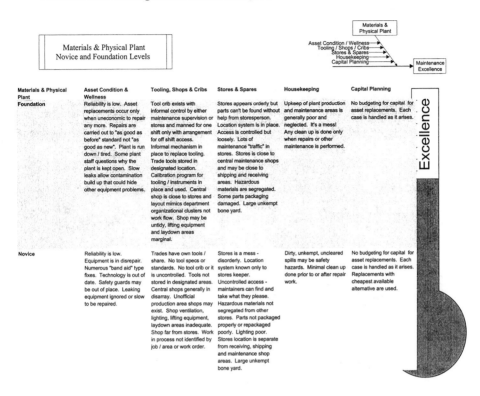

Figure 2-6   Materials and physical plant (the result): basic levels.

changing what maintenance is doing, to strengthen choices that will optimize business objectives.

Methodologies are the systematic methods or procedures used to apply logic principles. The broad category of continuous improvement includes several maintenance methodologies covered in depth in this book:

- Reliability-Centered Maintenance (Chapter 7). RCM focuses on overall equipment reliability. It appears at the journeyman and mastery levels in Figure 2-3, under Reliability Management.

Materials & Physical Plant
Apprentice, Journeyman, Mastery
Levels

Materials &
Physical Plant

Asset Condition / Wellness
Tooling / Shops / Cribs
Stores & Spares
Housekeeping
Capital Planning

Maintenance
Excellence

Excellence

Materials & Physical Plant	Asset Condition & Wellness	Tooling, Shops & Cribs	Stores & Spares	Housekeeping	Capital Planning
Mastery	Plant equipment is considered reliable, looks close to "new" even if old. Technology has been upgraded. Cleanliness of production equipment reveals sense of ownership.	Purpose built area centers. Tool crib attendant repairs tooling. Bays for fleet repair work by type of work with dedicated crews. Remote PLC programming. Separate laydown for work in process, materials receiving and outbound completed work.	Access to stores is controlled but open to maintainers. Staging areas for pre-kitting of all parts ordered in and from stock for specific work orders / projects prior to delivery to shops / maintainers. Large capital spares located near point of use or in special storage.	Well kept and clean plant production, maintenance and office areas. Area teams take pride in equipment upkeep. Cleaning is seen as an effective tool in keeping equipment in good condition.	Multi-year long range asset replacement strategy used for capital planning. Annual budget process refines long range plan.
Journeyman	Reliability is good but improvement is being sought. Plant appears to be in good operating condition. Asset replacement being done as needs arise with a one year look ahead at equipment condition as input to replacement decisions.	Area shops used. Shops adjacent production areas. Central shops used to support area teams for jobs too large to handle in area. Fleet shops purpose designed. Area and central shops all have on-line access to maintenance management system. Tools in secure area adjacent shop / bay. Tool crib for special tools. Sign out system for tooling.	Access to stores is controlled but open to maintainers. Areas for pre-kitting of parts ordered in for specific work orders / projects. Little "dust" indicates few obsolete items. Bone yard is used only for oversized, weather proof items.	Production and maintenance areas receive daily clean up. Cleaning is still a "chore" and carried out to medium standards. Dedicated clean up crews may exist. Special cleaning routines used on a weekly or monthly basis to bring areas back to good condition.	One year look ahead for capital budget needs supported by equipment condition assessments. Capital expenditure history used to forecast replacement funding for equipment upgrades. Betterment projects treated as stand alone projects.
Apprentice	Reliability is improving quickly in the plant. Plant appears to be a bit run down but generally operable. Asset replacement considered only if need is evident and trigged by annual budget cycle. Repairs carried out to "as good as new" standard.	Formal tool replacement mechanism and tool crib manned on all shifts, controlled and orderly. Tools tracked as stores items. Central shops designated for each trade or production area. Shop layout follows material flow. Shops well lit and ventilated. Area shops (if any) exist where space available. On-line access to maintenance management system in shops.	"Ready use" or "pre-expended" high usage low value stock available in shop areas. Stores traffic is down and access is well controlled. Stores is orderly and parts can be found quickly. Shops, shipping and receiving areas handy to stores. Separate areas for quarantine items, warranty items, receiving and shipping and repairable awaiting work. Bone yard is catalogued and orderly.	Weekly clean up of production and maintenance areas used. Standards of cleanliness are good but enforced only in cyclical clean ups.	Asset replacements identified only when budget cycle calls for estimates.

Figure 2-7   Materials and physical plant (the result): top levels.

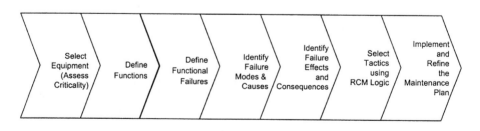

Figure 2-8   The basic steps of RCM.

- Root Cause Failure Analysis (see next section).
- Total Productive Maintenance implementation (Chapter 8). TPM focuses on achieving high reliability from operators and maintainers.
- Optimizing maintenance and materials management decisions (the entire book).

2.5.1 RCM

Reliability-Centered Maintenance aims to achieve maximum system reliability using maintenance tactics that can be effectively applied to specific system failures in the operating environment. The RCM process uses equipment and system knowledge to decide which maintenance interventions to use for each failure mode. That knowledge includes:

- System diagrams and drawings
- Equipment manuals
- Operational and maintenance experience with the system
- Effects of individual failures on the system, the entire operation, and its operating environment.

The RCM basic steps are listed in Figure 2-8, which shows that RCM results in a maintenance plan. The various decisions made for each failure are put into logical groupings for detailed planning, scheduling, and execution, as part of the overall maintenance workload. For each task, it is stated what must be done to prevent, predict, or find specific failures, and there is a specified frequency. Optimization techniques are used to determine the best frequency for each task and decide on corrective actions. This is based on monitoring results from the planned maintenance program, covered in Chapters 10, 11, and 12.

2.5.2 Root Cause Failure Analysis

Root Cause Failure Analysis (RCFA) is one of the basic reliability-enhancement methods. It appears at the journeyman level of Reliability Management in Figure 2-4. RCFA is relatively easy to perform, and many companies already do it—some using rigorous problem-solving techniques and some informally. Later in the chapter, you'll learn about a formal method based on easy-to-follow cause-and-effect logic, but first we look at informal and other problem-solving techniques. In all methods, the objective is to completely eliminate recurring equipment or system problems, or at least substantially diminish them.

Informal RCFA techniques are usually used by individuals or small groups to determine the best corrective action for a problem. Typically, this involves maintenance tradespeople, technicians, engineers, supervisors, superintendents, and managers. Drawing heavily on their own experience and information from such sources as trade periodicals, maintainers from other plants, and contractors, they often have immediate success. There are plenty of pitfalls, though, that can impair the informal approach:

- If only tradespeople do the RCFA, their solutions are often limited to repair techniques, parts and materials selection, and other design flaws.
- A restrictive engineering change control or spare parts (add to inventory) process can derail people who aren't skilled or accustomed to dealing with bureaucracy.
- If only senior staff do RCFA, they can miss out on technical details that the tradespeople would catch.
- Some organizations have a tendency to affix blame rather than fix the problem.

In short, informal techniques can work well but they have limitations and it can be hard to develop long-term solutions. All RCFA techniques face the same challenges, but they're greater if the process isn't formalized in some way.

More formal problem-solving techniques can be used very effectively. Consulting and educational organizations teach several techniques, two of which we examine here.

2.5.2.1 *What, Where, When Problem Solving*

The first problem-solving process is relatively straightforward:

- Establish the problem, noting what has actually changed from "normal" to unacceptable.
- Describe the problem, asking *what, where,* and *when* questions to determine the extent of it. Quantify what went wrong and be specific, so that you solve the problem only where it exists. You need to understand what is and is not happening now, as well as where and when.
- Identify possible causes.
- Identify the most likely cause. Test these possible causes against the IS and IS NOT criteria for the "what, when, and where" of the problem statement.
- Verify the cause. Test any assumptions you have made, looking for holes in the argument.
- Implement a solution that addresses the cause.

These formal problem-solving techniques, usually performed in a structured group with a facilitator, are very effective. To use this approach, set up a weekly or monthly meeting to identify and prioritize problems that need solutions. Although day-to-day maintenance and equipment issues will figure prominently, your mandate will likely extend beyond them.

Problem-solving groups should include a cross-section of interested stakeholders such as production, finance, human resources, training, and safety representatives as well as maintenance. Because its purview is so broad, the group is often most effective broken into smaller task teams. Each is assigned a specific problem to analyze, and usually made responsible for solving it completely. These teams report to the problem-solving group on progress and solutions.

Formal problem-solving groups usually produce credible results because of their broad representation and rigorous formal analytical processes. This thorough approach ensures solutions that work over the long term.

2.5.2.2 Cause and Effect

The second RCFA technique we examine is based on cause-and-effect logic. Theoretically, all events are the result of potentially infinite combinations of pre-existing conditions and triggering events. Events occur in sequence, with each event triggering other events, some being the failures that we are trying to eliminate.

Think of these sequences as chains of events, which are only as strong as their weakest link. Break that link and the chain fails—the subsequent events are changed. Even if you eliminate the failure you're targeting, though, you could also trigger some other chain of events. Remember that the solution to one problem may well turn out to be the cause of another.

To perform a cause-and-effect RCFA, you need to:

- Identify the unacceptable performance
- Specify what is unacceptable (like the what, when and where of the previous method)
- Ask "what is happening?" and "what conditions must exist for this event to happen?"
- Continue to ask this combination of "what" questions until you identify some event that can be controlled. If that event can be changed to prevent the failure reoccurring, you have a "root cause" that can be addressed
- Eliminate the "root cause" through an appropriate change in materials, processes, people, systems or equipment.

By repeating the *what* questions, you usually get a solution within five to seven iterations. A variation of this process asks *why* instead of *what*—both questions work.

Because it is performed formally, with a cross-section of stakeholders exercising complete control over the solution, the success rate of cause and effect is also high.

2.6 OPTIMIZING MAINTENANCE DECISIONS: BEYOND RCM

Managing maintenance goes beyond repair and prevention to encompass the entire asset life cycle from selection to disposal. Key life-cycle decisions that must be made include:

- Component replacements
- Capital equipment replacements
- Inspection result decisions
- Resource requirements

To make the best choices, you need to consider not only technical aspects but historical maintenance data, cost information, and sensitivity testing to ensure you meet your objectives in the long run. Jardine (3) describes several situations that can be dealt with effectively:

- Replacements when operating costs increase with use
- Replacements when operating costs increase with time
- Replacement of a machine when it is in standby mode
- Capital equipment replacements to maximize present value
- Capital equipment replacements to minimize total cost
- Capital equipment replacements considering technology improvements over time
- Optimizing replacement intervals for preventive component replacement

- Optimizing replacement intervals that minimize downtime
- Group replacements
- Optimizing inspection frequencies to maximize profit or to minimize downtime
- Optimizing inspection frequencies to maximize availability
- Optimizing inspection frequencies to minimize total costs
- Optimizing overhaul policies
- Replacement of monitored equipment based on inspection, cost, and history data

All these methods require accurate maintenance history data that shows what happened and when. Specifically, you need to make the distinction between repairs due to failures and ones that occurred while doing some other work. Note that it is the quality of the information that is important, not the quantity. Some of these decisions can be made with relatively little information, as long as it is accurate.

RCM produces a maintenance plan that defines what to do and when. It is based on specific failure modes that are either anticipated or known to occur. Frequency decisions are often made with relatively little data or a lot of uncertainty. You can make yours more certain and precise by using the methods described later in the book, and in Jardine's work (3). Along with failure history and cost information, these decision-making methods show you how to get the most from condition-monitoring inspections, to make optimal replacement choices.

All these decisions can be made with sufficient data and, in several cases, computerized tools designed for the purpose. These currently tend to be standalone tools used primarily by engineers and highly trained technicians. At the time of writing, combining them with computerized maintenance management and condition-based monitoring systems is being explored. Maintenance management systems are growing beyond managing activities and the transactions around them, to becoming management and decision support systems. Eventually, this will make the job of monitoring personnel much easier. For more information, see Chapter 4, on data acquisition, and Chapter 10, on modeling.

REFERENCES

1. John Dixon Campbell. Uptime: Strategies for Excellence in Maintenance Management. Portland, OR: Productivity Press, 1995.
2. John P. Kotter. Leading Change. Boston: Harvard Business School Press, 1996.
3. A. K. S. Jardine. Maintenance, Replacement and Reliability. Pitman/Wiley, 1973.

3

Measurement in Maintenance Management

Performance management is one of the basic requirements of an effective operation. Doing it well, though, isn't as straightforward as it may seem. In this chapter, we discuss effective tools to help you strengthen your maintenance performance measurement.

To start, we summarize measurement basics with an example that shows how numeric measures can be both useful and misleading. Measurement is important in maintenance continuous improvement and in identifying and resolving conflicting priorities. We explore this within the maintenance department, and between Maintenance and the rest of the organization. We look at both macro and micro approaches, using a variety of examples. At the macro level, this includes the Balanced Scorecard; at the micro level, shaft-alignment case history.

Maintenance performance measurement is then subdivided into its five main components: productivity, organization, work efficiency, cost, and quality, together with some overall measurements of departmental results. You'll learn about consistency and reliability as they apply to measurement. A major section of the chapter is devoted to individual performance measures, with sample data attached. We summarize the data required to complete these measures, for you to decide whether you can use them in your own workplace. We also cover the essential tie-in between performance measures and action.

The chapter concludes with a practical look at using the benefits/difficulty matrix as a tool for prioritizing actions. You will also find a useful step-by-step guide to implementing performance measures.

3.1   MAINTENANCE ANALYSIS: THE WAY
INTO THE FUTURE

There is no mystique about performance measurement. The trick is how to use the results to achieve the needed actions. This requires several conditions: consistent and reliable data, high-quality analysis, clear and persuasive presentation of the information, and a receptive work environment.

Since maintenance optimization is targeted at executive management and the boardroom, it is vital that the results reflect the basic business equation:

Maintenance is a business process
turning inputs into useable outputs.

Figure 3-1 shows the three major elements of this equation: the inputs, the outputs, and the conversion process. Most of the inputs are familiar to the maintenance department and

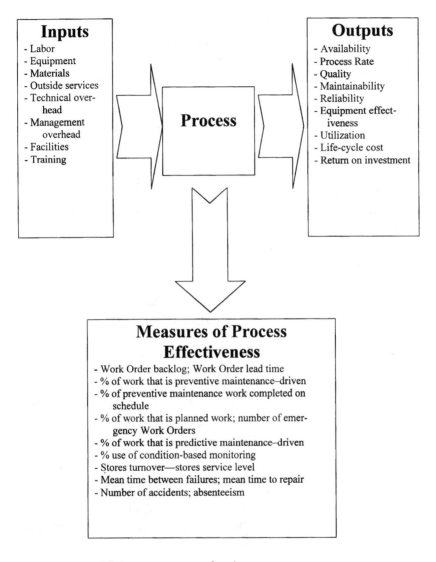

Figure 3-1 Maintenance as a business process.

readily measured, such as labor costs, materials, equipment, and contractors. There are also inputs that are more difficult to measure accurately—including experience, techniques, teamwork, and work history—yet each can significantly impact results.

Likewise, some outputs are easily recognized and measured, others are harder. As with the inputs, some are intangible, like the team spirit that comes from completing a difficult task on schedule. Measuring attendance and absenteeism isn't exact and is no substitute for these intangibles, and overall performance indicators are much too high-level. Neither is a substitute for intangible benefits. Although intangibles contribute significantly to overall maintenance performance, the focus of this book is on the tangible measurements.

Converting the maintenance inputs into the required outputs is the core of the maintenance manager's job. Yet rarely is the absolute conversion rate of much interest in itself. Converting labor hours consumed into reliability, for example, makes little or no sense—until it can be used as a comparative measure, through time or with a similar division or company. Similarly, the average material consumption per Work Order isn't significant—until you see that Press #1 consumes twice as much repair material as Press #2 for the same production throughput. A simple way to reduce materials per Work Order consumption is to split the jobs, increasing the number of Work Orders (this doesn't do anything for productivity improvement, of course).

The focus, then, must be on the comparative standing of your company or division, or improving maintenance effectiveness from one year to the next. These comparisons highlight another outstanding value of maintenance measurement—it regularly compares progress toward specific goals and targets. This benchmarking process—through time or with other divisions or companies—is increasingly being used by senior management as a key indicator of good maintenance management. It often discloses surprising discrepancies in performance. A recent benchmarking exercise turned up the data from the pulp industry shown in Table 3-1.

The results show some significant discrepancies, not only in the overall cost structure but in the way Company X does business. It has a heavy management structure and hardly uses outside contractors, for example. You can clearly see from this high-level benchmarking that, in order to preserve Company X's competitiveness in the marketplace, something needs to be done. Exactly what, though, isn't obvious. This requires more detailed analysis.

As you'll see later, the number of potential performance measures far exceeds the maintenance manager's ability to collect, analyze, and act on the data. An important part, therefore, of any performance measurement implementation is to thoroughly understand the few, key performance drivers. Maximum leverage should always take top priority. First, identify the indicators that show results and progress in areas that most critically need improvement. As a place to start, consider Figure 3-2.

If the business could sell more products or services with a lower price, it is cost-constrained. The maximum payoff is likely to come from concentrating on controlling inputs, i.e., labor, materials, contractor costs, overheads. If the business

Table 3-1   Data from the Pulp Industry

	Average	Company X
Maintenance costs—$ per ton output	78	98
Maintenance costs—$ per unit equipment	8900	12700
Maintenance costs as % of asset value	2.2	2.5
Maintenance management costs as % of total maintenance costs	11.7	14.2
Contractor costs as % of total maintenance costs	20	4
Materials costs as % of total maintenance costs	45	49
Total number of Work Orders per year	6600	7100

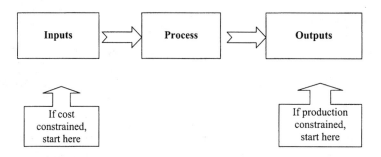

Figure 3-2   Maintenance optimizing: where to start.

can profitably sell all it produces, it is production-constrained.
It's likely to achieve the greatest payoff from maximizing outputs through asset reliability, availability, and maintainability.

3.1.1   Keeping Maintenance in Context

As an essential part of your organization, maintenance must
adhere to the company's overall objectives and direction.
Maintenance cannot operate in isolation. The continuous improvement loop (Figure 3-3), key to enhancing maintenance,
must be driven by and mesh with the corporation's planning,
execution, and feedback cycle.

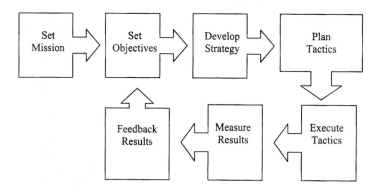

Figure 3-3   The maintenance continuous-improvement loop.

Disconnects frequently occur when the corporate and department levels aren't in synch. For example, if the company places a moratorium on new capital expenditures, this must be fed into the equipment maintenance and replacement strategy. Likewise, if the corporate mission is to produce the highest-quality product possible, this probably doesn't correlate with the maintenance department's cost-minimization target. This type of disconnect frequently happens inside the maintenance department itself. If your mission is to be the best-performing maintenance department in the business, your strategy must include condition-based maintenance and reliability. Similarly, if the strategy statement calls for a 10% reliability increase, reliable and consistent data must be available to make the comparisons.

3.1.2 Conflicting Priorities for the Maintenance Manager

In modern industry, all maintenance departments face the same dilemma: which of the many priorities to put at the top of the list? Should the organization minimize maintenance costs or maximize production throughput? Does it minimize downtime or concentrate on customer satisfaction? Should it spend short-term money on a reliability program to reduce long-term costs?

Corporate priorities are set by the senior executive and ratified by the Board of Directors. These priorities should then flow down to all parts of the organization. The Maintenance Manager must adopt those priorities; convert them into corresponding maintenance priorities, strategies, and tactics to achieve the results; then track them and improve on them.

Figure 3-4 is an example of how corporate priorities can flow down through the maintenance priorities and strategies to the tactics that control the everyday work of the maintenance department. If the corporate priority is to maximize product sales, maintenance can focus on maximizing throughput and equipment reliability. In turn, the mainte-

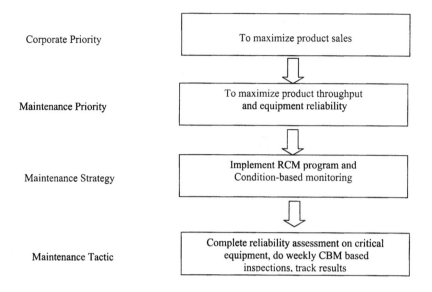

Corporate Priority — To maximize product sales

Maintenance Priority — To maximize product throughput and equipment reliability

Maintenance Strategy — Implement RCM program and Condition-based monitoring

Maintenance Tactic — Complete reliability assessment on critical equipment, do weekly CBM based inspections, track results

Figure 3-4 Interrelating corporate priorities with maintenance tactics.

nance strategies will also reflect this, and could include, for example, implementing a formal reliability-enhancement program supported by condition-based monitoring. Out of these strategies flow the daily, weekly, and monthly tactics. These, in turn, provide lists of individual tasks that then become the jobs that will appear on the Work Orders from the Enterprise Asset Management System (EAMS) or Computerized Maintenance Management System (CMMS). Using the Work Order to ensure that the inspections get done is widespread. Where organizations frequently fail is in completing the follow-up analysis and reporting on a regular and timely basis. The most effective method is to set them up as weekly Work Order tasks, subject to the same performance tracking as preventive and repair Work Orders.

Trying to improve performance, you can be confronted with many, seemingly conflicting, alternatives. Numerous review techniques are available to establish how your organization compares to industry standards or best maintenance practice. The most effective techniques help you map priorities by indicating the payoff that improvements will make. The review techniques tend to be split into macro (covering the full maintenance department and its relationship to the business) and micro (with the focus on a specific piece of equipment or a single aspect of the maintenance function).

The leading macro techniques are:

- Maintenance effectiveness review—involves the overall effectiveness of maintenance and its relationship to the organization's business strategies. These can be conducted internally or externally, and typically cover areas such as:
  Maintenance strategy and communication
  Maintenance organization
  Human resources and employee empowerment
  Use of maintenance tactics
  Use of reliability engineering and reliability-based approaches to equipment performance monitoring and improvement
  Information technology and management systems
  Use and effectiveness of planning and scheduling
  Materials management in support of maintenance operations
- External benchmark—draws parallels with other organizations to establish how the organization compares to industry standards. Confidentiality is a key factor, and results typically show how the organization ranks within a range of performance indicators. Some of the areas covered in benchmarking overlap with the maintenance effectiveness review. Additional topics include:
  Nature of business operations
  Current maintenance strategies and practices
  Planning and scheduling

Inventory and stores management practices
Budgeting and costing
Maintenance performance and measurement
Use of CMMS and other IS (information system) tools
Maintenance process re-engineering

- Internal comparisons—measure a set of parameters similar to those of the external benchmark, but draw from different departments or plants. They are generally less expensive and, if the data is consistent, illustrate differences in maintenance practices among similar plants. From this, you can decide which best practices to adopt.
- Best Practices Review—looks at maintenance's process and operating standards and compares them against the industry best. This is generally the starting point for a maintenance process upgrade program, focusing on areas such as:
Preventive maintenance
Inventory and purchasing
Maintenance workflow
Operations involvement
Predictive maintenance
Reliability-Based Maintenance
Total Productive Maintenance
Financial optimization
Continuous improvement
- Overall Equipment Effectiveness (OEE)—measures a plant's overall operating effectiveness after deducting losses due to scheduled and unscheduled downtime, equipment performance, and quality. In each case, the subcomponents are meticulously defined, providing one of the few reasonably objective and widely used equipment-performance indicators.

Table 3-2 summarizes one company's results. Remember that the individual category results are multiplied through the calculation to derive the final result. Although Company Y achieves 90% or higher in

Table 3-2 Example of Overall Equipment Effectiveness

	Target (%)	Company Y (%)
Availability	97	90
×		
Utilization rate	97	92
×		
Process efficiency	97	95
×		
Quality	99	94
=		
Overall equipment effectiveness	90	74

each category, it will have an OEE of only 74% (see Chapter 8 for further details of OEE). This means that by increasing the OEE to, say, 95%, Company Y can increase its production by (95 − 74)/74 = 28% with minimal capital expenditure. If you can accomplish this in three plants, you won't need to build a fourth.

These, then, are some of the high-level indicators of the effectiveness and comparative standing of the maintenance department. They highlight the key issues at the executive level, but more detailed evaluation is needed to generate specific actions. They also typically require senior management support and corporate funding—not always a given.

Fortunately, there are many maintenance measures that can be implemented that don't require external approval or corporate funding. These are important because they stimulate a climate of improvement and progress. Some of the many indicators at the micro level are:

- Post implementation review of systems to assess the results of buying and implementing a system (or equipment) against the planned results or initial cost justification
- Machine reliability analysis/failure rates—targeted at individual machine or production lines

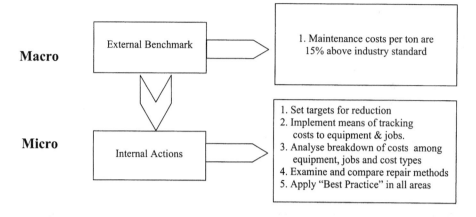

Figure 3-5   Relating macro measurements to micro tasks.

- Labor effectiveness review—measuring staff allocation to jobs or categories of jobs compared to last year
- Analyses of materials usage, equipment availability,
  utilization, productivity, losses, costs, etc.

All these indicators give useful information about the maintenance business and how well its tasks are being performed.
You have to select those that most directly achieve the maintenance department's goals as well as those of the overall business.

Moving from macro or broad-scale measurement and optimization to a micro model can create problems for Maintenance Managers. You can resolve this by regarding the macro
approach as a project or program and the micro indicators
as individual tasks or series of tasks. The example in Figure
3-5 shows how an external benchmark finding can be translated into a series of actions that can be readily implemented.

3.2   MEASURING MAINTENANCE:
THE BROAD STROKES

To improve maintenance management, you need measurement capability for all the major items under review. How-

ever, as previously mentioned, there usually aren't enough resources to go beyond a relatively small number of key indicators. The following are the major categories that should be considered:

- Maintenance productivity—measures the effectiveness of resource use
- Maintenance organization—measures the effectiveness of the organization and planning activities
- Efficiency of maintenance work—how well maintenance keeps up with workload
- Maintenance costs—overall maintenance cost vs. production cost
- Maintenance quality—how well the work is performed
- Overall maintenance results—measures overall results

Measurements are only as good as the actions they prompt—the results are as important as the numbers themselves. Attractive, well-thought-out graphics will help to "sell" the results and stimulate action. Graphic layouts should be informative and easy to interpret, like the spider diagram in Figure 3-6. Here, the key indicators are measured on the radial arms in percentage achievements of a given target. Each measurement's goals and targets are also shown, with the performance gap clearly identified. Where the gap is largest, the shortfall is greatest—and so is the need for immediate action.

The spider diagram in Figure 3-6 shows the situation at a point in time, but not its progress through time. For that, you need trend lines. These are best shown as a graph (or series of graphs) with the actual status and the targets clearly identified, as in Figure 3-7. For a trend line to be effective, the results must be readily quantifiable and directly reflect the team's efforts.

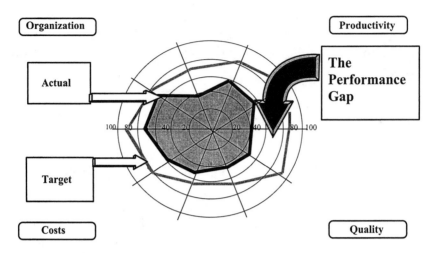

Figure 3-6 Spider diagram showing performance gaps.

3.2.1 Balanced Scorecard

Each of the examples above shows only a single measure of performance. The Balanced Scorecard concept broadens the measure beyond the single item. Each organization should develop its own Balanced Scorecard to reflect what motivates its business behavior.

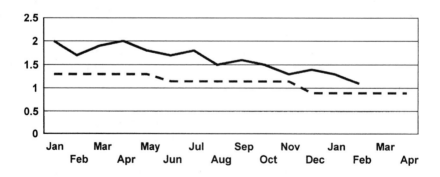

Figure 3-7 Trend-line of quality results.—rework per 100 jobs; ——— target.

Table 3-3 shows an example combining the elements of the input–process–output equation referred to earlier, with leading and lagging indicators, and short- and long-term measurements:

- Leading indicators—the change in the measurement (hours of training, for example) precedes the improvements being sought (decreases in error rates). Typically, you see these benefits only at a later date.

Table 3-3   Example of a Balanced Scorecard

Inputs	Money spent on
	Labor
	Materials
	Services
Process	Backlog of Work Orders
	Compliance of actual work with planned
	Work distribution among shifts or trades
	Rework required on Work Orders
Outputs	Reliability
	Availability
	Maintainability
	OEE
Short-term	Number of breakdowns
	Number of on-time Work Order completions
	Number of stock-outs
	Number of daily Work Orders
Long-term	Maintenance costs as % of replacement asset value
	MRO inventory turns
	Number of PMs reviewed
	Organizational levels
Leading indicators	Number of training hours
	% of Work Orders driven by CBM
	RCM use
	Parts rationalization
Lagging indicators	Amount of time lost through injuries and absences
	Total maintenance cost variance from budget
	Emergency WOs as % of total WOs
	Staff turnover

- Lagging indicators—the change in the measurement (staff quitting, for example) lags behind the actions that caused it (overwork or unappreciative boss).

For each of these elements, four representative indicators were developed. Although each indicator shows meaningful information, the Balanced Scorecard provides a good overview of the effectiveness of the total maintenance organization (Table 3-3). In a later section of this chapter, we examine some of the pros and cons of this increasingly popular measurement technique.

The broad performance measures are essential to understand the overall direction and progress of the maintenance function. But within this broad sweep lie multiple opportunities to measure small but significant changes in equipment operation, labor productivity, contractors' performance, material use, and the contribution of technology and management. The next section examines some of these changes and provides examples that can be used in the workplace.

3.3 MEASURING MAINTENANCE: THE FINE STROKES

To understand individual elements of maintenance functions, you need analysis at a much more detailed level. To evaluate, predict, and improve the performance of a specific machine, you must have its operating condition and repair data, not only for the current period but also historically. Also, it's useful to compare data from similar machines. Later in this chapter, we examine the sources of this data in more detail. For now, you should know that the best data sources are the CMMS, EAM, CBM, SCADA (Supervisory Control and Data Acquisition Systems), and Process Control systems currently in widespread use.

Among the many ways to track individual equipment performance are the measurements of reliability, availability, productivity, life-cycle costs, and production losses. Use these techniques to identify problems and their causes so that remedial action can be taken. An interesting case study examined a series of high-volume pumps to establish why the running costs (i.e., operating and regular maintenance costs) varied so widely among similar models. Shaft alignment proved to be the major problem, setting off improvements that drove the annual cost per unit down from over $30,000 to under $5000, as shown in Figure 3-8.

Despite this remarkable achievement, the company didn't get the expected overall benefit. After further special analysis, two additional problems were found. First, the O&M costs excluded contractor fees and so should have been labeled "Internal O&M Costs." Then, the subcontractor's incremental cost to reduce the vibration from the industry standard of 0.04"/sec to the target level of 0.01"/sec (see Figure 3-9) was unexpectedly higher than predicted from earlier improvements.

When the O&M costs were revised to include the subcontractor fee, the extra effort to move from 0.04"/sec to 0.01"/sec was at a considerably higher cost than the savings in the operating costs. The optimal position for the company was to maintain at 0.04"/sec (see Figure 3-10).

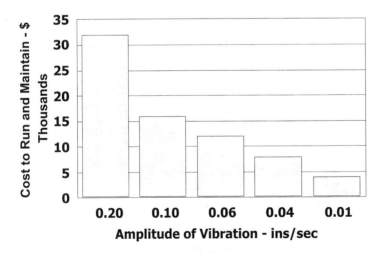

Figure 3-8 Internal pump operating and maintenance costs.

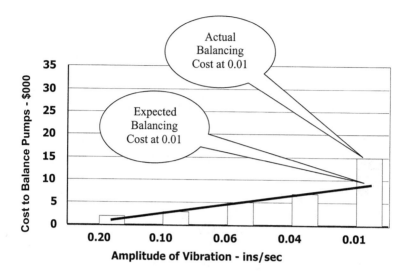

Figure 3-9   Subcontractor cost of pump balancing.

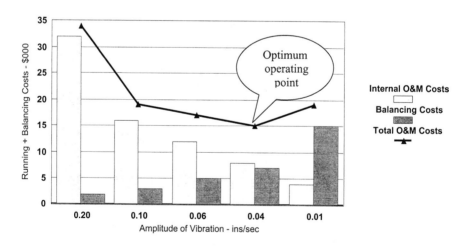

Figure 3-10   Total pump O&M costs.

This type of quasi-forensic analysis can produce dramatic savings. In the case summarized above, the company had 50 of these pumps operating and was able to reduce costs from $1.6 M to $750,000, saving 53%. By taking the final step to refine the balancing from 0.04 to 0.01, the company spent an extra $200,000 for no additional tangible benefit.

You can use similar investigative techniques to evaluate and improve labor and materials consumption. Variable labor consumption among similar jobs at a number of plants can identify different maintenance methodologies and skill levels. By adopting the best practices and adding some targeted training, you can achieve significant improvements and avoid sizeable problems. In one example, a truck motor overheating was traced back through the CMMS Work Order to a badly seated filter. The maintenance technician responsible for the work had insufficient training for the task. Emergency recalls were issued for six other trucks on the highway that he had fitted the same way, and each one was fixed without damage. The technician received additional training and got to keep his job.

Variations in materials usage can be tracked from the Work Orders and the inventory records, leading to standardized methods. More recently, multiplant operations have been able to access data from multiple databases for analytical and comparative purposes. Parts specification has become standardized, creating huge savings through bulk buying and centralized storage. There are several examples in which a combination of reduced inventory, reduced supplier base, better negotiated prices, and the removal of many hidden purchasing costs through improved productivity has generated savings well above 15% in the inbound supply chain.

In this chapter, we have introduced a wide variety of topics to set the scene for later and more detailed examination. The core issues are the same throughout this book—what should the Maintenance Manager optimize and how should it be done?

3.4 WHAT'S THE POINT OF MEASURING?

Organizations strive to be successful, to be considered the best or the industry leader. In an earlier section of this chapter, we reviewed how objectives cascade from the corporate level to the maintenance department and get translated into actionable tactics. It's assumed through this process that there is a common set of consistent and reliable performance measures that "prove" that A is better than B. This is actually not the case, but through the work of Coetzee (1), Wireman (2), MIMOSA (3), and others, common standards are starting to emerge.

A major driving force for performance measurement is the goal of achieving excellence. There must be effective measurement methods to withstand the scrutiny of the Board of Directors, shareholders, and senior management. As the mission statement cascades down to the maintenance department, so does the demand for accurate measurement. This is reason enough to measure maintenance performance.

There are many other reasons, though, to make improvements, including:

- Competitiveness—regardless of whether the goals are price-, quality-, or service-driven, you must compare to establish how competitive you are.
- Right-sizing, down-sizing, and up-sizing—adjusting the size of the organization to deliver products and services while continuing to prosper becomes meaningless if you can't realistically measure performance.
- New processes and technologies are being introduced rapidly—not only in manufacturing but also in maintenance. To produce the expected improvements, you must keep careful track of the results.
- Performance measurement is integral when deciding to maintain or replace an item. An example of life-

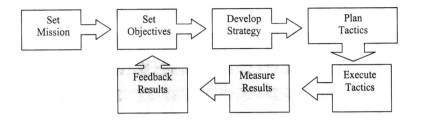

Figure 3-11 Measurements as a core part of the performance-improvement loop.

cycle costing is used later in this chapter to determine whether to maintain or replace.

- The performance-improvement loop (Figure 3-11) is the core process in identifying and implementing progress. Performance measurement and results feedback are essential elements in this loop.

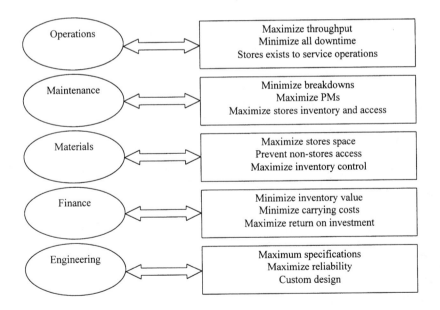

Figure 3-12 Conflicting departmental objectives.

Table 3-4 A Hierarchical Approach to Performance Indicators

Area covered by indicators	Functional areas	No. of indicators	Sample indicators
Corporate	N/A	4	Return on net assets; total cost to produce
Financial	N/A	8	Maintenance cost per unit produced; replacement value of assets maintained
Efficiency and effectiveness	Preventive maintenance	6	% of total direct maintenance cost that is break-down-related
	Work Order systems	3	% total WOs that are PMs
	Training	4	% of total maintenance work caused by rework due to lack of skills
	Operational in-volvement	2	Maintenance-related equipment downtime this year versus last year
	Predictive maintenance	1	Current maintenance costs versus those prior to predictive program
	Reliability-Centered Maintenance	7	Number of repetitive failures versus total failures
	Total Productive Main-tenance	4	Overall equipment effectiveness combining availabil-ity, performance efficiency, and quality rate
Tactical	Preventive maintenance	2	% of total number of breakdowns that should have been prevented
	Inventory and procure-ments	4	Total of items filled on demand versus total re-quested
	Work Order systems	4	Total planned WOs versus total WOs received
	CMMS	4	Total costs charged to equipment versus total costs from Accounting
	Operational in-volvement	3	PM hours performed by operators as % of total maintenance hours
	Reliability-Centered Maintenance	2	Number of equipment breakdowns per hour oper-ated

Functional		
Preventive maintenance	3	% of total WOs generated from PM inspections
Inventory and procurements	4	% of total stock items inactive
Work Order systems	7	% of total labor costs from WOs
Planning and scheduling	2	% of total labor costs that are planned
CMMS	6	% of total in plant equipment in CMMS
Training	5	Training hours per employee
Operational involvement	5	% of total hours worked by operators spent on equipment improvement
Predictive maintenance	2	PdM hours % of total maintenance hours
Reliability-Centered Maintenance	3	% of failures where root cause analysis is performed
Total Productive Maintenance	2	% of critical equipment covered by design studies
Statistical financial optimization	3	% of critical equipment where maintenance tasks are audited
Continuous improvement	3	Savings from employee suggestions

3.5 WHAT SHOULD WE MEASURE?

In the next section, we look at a wide variety of performance measurements, with sources for further study. You must concentrate on the need for consistency and reliability. Comparisons over time and between equipment or departments must be consistent to be valid. You want the assessment of your operation to be reliably complete, without significant omissions.

To start, you must understand that maintenance management is a dynamic process, not static. It is inextricably linked to business strategy, not simply a service on demand. Finally, it is an essential part of the business process, not just a functional silo operating in isolation.

Despite this, conflicts frequently arise in setting objectives for an organization. For example, the objectives in Figure 3-12, taken from a strategic review of Company C, will be confusing when translated into performance measurement, and later action.

From the discussion so far, you can see that there is no real consensus on the precise source data, what should be measured, and how it should be analyzed. Wireman (2) takes a hierarchical approach in his book *Developing Performance Indicators for Managing Maintenance*. With comprehensive coverage, he develops performance measures based on a five-tier hierarchy: corporate, financial, efficiency and effectiveness, tactical, and functional. He covers the areas shown in Table 3-4. With over 100 indicators listed, Wireman clearly shows that you can measure just about anything. He also emphasizes the equal and opposite need to be selective about what is measured.

In his book *Maintenance* (1), Coetzee's approach centers on the four pillars of results, productivity, operational purposefulness, and cost justification. The following approach expands his method by adding some extra measures and presenting them as calculations. Check the core data set in

Section 3.7. We emphasize that new measurement formulae are being continually developed; those presented here are examples only, not to be considered a complete or recommended set. What makes the best set of data will vary considerably with each situation.

3.6 MEASURING OVERALL MAINTENANCE PERFORMANCE

The measures here are macro-level, showing progress toward achieving the maintenance department's overall goals. Later in this chapter, we cover micro measurement that applies to individual equipment. Figure 3-13 summarizes the categories we use in the examples.

3.6.1 Overall Maintenance Results

These indicators measure whether the maintenance department keeps the equipment productive and produces quality product. We look at the following five measures.

1. *Availability* is the percentage of time that equipment is available for production, after all scheduled and unscheduled downtime. Note that idle time caused by lack of product demand isn't deducted from the total time available. The

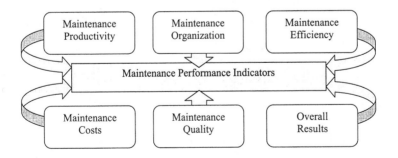

Figure 3-13 Maintenance categories for macro analysis.

equipment is considered "available" even though no production is demanded.

$$\text{Availability} = \frac{\text{Total time} - \text{Downtime}}{\text{Total time}}$$

Downtime includes all scheduled and unscheduled downtime, but not idle time through lack of demand.

Total time = 8760 hours

Downtime = 392 hours

$$\text{Availability} = \frac{8760 - 392}{8760} = 95.5\%$$

2. *Mean Time to Failure* (MTTF) is a popular measure that will be revisited in some depth later in this book. It represents how long a machine can be expected to run before it dies. It measures average uptime, and is widely used in production scheduling to determine whether the next batch can likely be produced without interruption. In this example, the equipment is expected to run an average of 647 hours from the previous failure.

$$\text{MTTF} = \frac{\text{Total time} - \text{Downtime} - \text{Nonutilized time}}{\text{Number of breakdowns}}$$

Total time = 8760 hours

Downtime = 392 hours

Nonutilized time = 600 hours

Number of breakdowns = 12

$$\text{MTTF} = \frac{8760 - 392 - 600}{12} = 647 \text{ hours}$$

3. *Failure frequency or breakdown frequency* measures how often the equipment is expected to fail. It is typically used as a comparative measure, not an absolute, and therefore

should be trended. It helps to regard it as the conditional probability of failure within the next time period. Using this measure, it should be note that "failure" is generally an inexact term. For example, if a machine that is designed to run at 100 bits per minute is running at 85, is this deemed "failure"? Similarly, if it produces at 100 bpm but 10 units are defective, is this "failure"? Adopt the Reliability-Centered Maintenance approach—the run rate and the quality rates are given quantifiable failure levels. Anything below is deemed to have failed, even though it may operationally still be struggling on.

Breakdown frequency

$$= \frac{\text{Number of breakdowns}}{\text{Totaltime} - \text{Downtime} - \text{Nonutilized time}}$$

Total time = 8760 hours

Downtime = 392 hours

Nonutilized time = 600 hours

Number of breakdowns = 12/year

Production rate = 10/hour

Breakdown frequency

$$= \frac{12}{8760 - 392 - 600} = 0.0015 \text{ Failures per hour}$$

= Probability of 0.15% of failure within the next hour

Probability of failure within the next production run

$$\text{of 2500 units} = 0.15\% \times \frac{2500}{10} = 37.5\%$$

4. *Mean Time to Repair* (MTTR)—the total time it takes to fix the problem and get the equipment operating again. This includes notification time, travel time, diagnosis time, fix time, wait time (for parts or cool down), reassembly time, and test time. It reflects how well the organization can respond to a problem and, from the list of the total time components, you can see that it covers areas outside the maintenance depart-

ment's direct control. MTTR also measures how long Operations will be out of production, broadly indicating the effect of maintenance on equipment production rate. Note that this can be used as a measure of the average of all MTTRs, as the MTTR for breakdowns, or scheduled outages.

$$\text{MTTR} = \frac{\text{Unscheduled downtime}}{\text{Number of breakdowns}}$$

Unscheduled downtime = 232 hours

Scheduled outages = 160

Number of breakdowns = 12

Number of scheduled outages = 6

MTTR for unscheduled downtime $= \dfrac{232}{12} = 19.3$ hours

MTTR for scheduled outages = 26.7 hours

MTTR for all downtime = 21.8 hours

5. *Production Rate Index*—impact of maintenance on equipment effectiveness. As with the previous indicator, you must interpret results carefully, because operating speeds and conditions will have an impact. To minimize the effect of these variations, it is trended over time as an index and has no value as an absolute number.

$$\text{Production Rate Index} = \frac{\text{Production rate (units/hour)}}{\text{Total time} - \text{Downtime} - \text{Nonutilized time}}$$

Production rate = 10 units/hour

Total time = 8760 hours

Downtime = 392 hours

Nonutilized time = 600 hours

$$\text{Production Rate Index} = \frac{10}{8760 - 392 - 600} = 0.001287$$

3.6.2 Maintenance Productivity

Maintenance Productivity Indices measure Maintenance's use of resources, including labor, materials, contractors, tools, and equipment. These components also form the cost indicators that will be dealt with later.

   1. *Manpower utilization* is usually called *wrench time*, because it measures the time consumed by actual maintenance tasks as a percentage of total maintenance time. The calculation includes standby time, wait time, sick time, vacation time, and time set aside for meetings, training, and so on. It measures only time spent on the job. Frequently there are problems measuring the results, because assigning time to jobs varies within organizations. For example, is travel time assigned to the job? For measurements within a single organization, the definitions need to be clearly defined, documented, and adhered to. In the following example, you'll see a wrench time figure of 69%—a fairly modest standard. High-performance, land-based, factory operations will exceed 80%.

$$\text{Manpower utilization} = \frac{\text{Wrench time}}{\text{Total time}}$$

32 staff, total time $= 32 \times 2088 = 66{,}816$

Wrench time $= 46{,}100$

$$\text{Manpower utilization} = \frac{46{,}100}{66{,}916} = 69\%$$

   2. *Manpower efficiency* shows the extent to which the maintenance jobs completed matched the time allotted during the planning process. Although typically called an "efficiency" measure, it is also a measure of the planning accuracy itself. Many EAM systems can modify the job planning times for repeat jobs based on the average of the past number of times the job has been completed. The manpower efficiency measure becomes a comparison with this moving average. Many planners reject this measure because they don't want to plan a new

job until the teardown and subsequent diagnosis have been completed. They'll apply it only to preventive maintenance or repeat jobs.

$$\text{Manpower efficiency} = \frac{\text{Time taken}}{\text{Planned time}}$$

Time taken = Wrench time = 46,100

Planned/allowed time = 44,700

$$\text{Manpower efficiency} = \frac{44,700}{46,100} = 97\%$$

3. *Materials usage per work order* measures how effectively the materials are being acquired and used. Again, this is a composite indicator that you must further refine before taking direct action. The measure shows the average materials consumption per Work Order. Variations from job to job can occur as a result of changes in buying practices, pricing or sourcing, inventory costing or accounting practices, the way the jobs are specified, parts replacement policy, and so on. As noted earlier, subdividing Work Orders greatly reduces the materials cost per Work Order. Nevertheless, it is a simple trend to plot and, as long as you haven't made significant underlying changes, indicates whether material usage is improving.

$$\text{Material usage} = \frac{\text{Total Materials cost charged to Work Orders}}{\text{Number of Work Orders}}$$

Total materials consumed = $1,400,000

Total WOs = 32,000

$$\text{Material usage} = \frac{1,400,000}{32,000} = \$44 \text{ per Work Order}$$

4. *Total maintenance costs as a percentage of total production costs* indicates the overall effectiveness of resource use. It suffers from the same variations as the material-usage measure shown above but, if you maintain consistent underlying policies and practices, it will show overall performance improvements or deterioration. Although this indicator is shown only for total costs, similar indices can be readily created for labor, contractors, and special equipment—in fact, any significant cost element.

Total maintenance costs = $4.0M

Total production costs = $45M

$$\text{Maintenance Cost Index} = \frac{4.0}{45} = 8.9\%$$

3.6.3 Maintenance Organization

Maintenance performance indicators measure the effectiveness of the organization and maintenance planning activities. This is frequently missed when considering the overall effectiveness of the department, because it takes place largely at the operational end. Studies have shown, though, that effective planning can significantly impact maintenance's operational effectiveness. In fact, one of the early selling features of the CMMS products was that allocating 5% of the maintenance department's work effort to planning would increase the overall group's efficiency by about 20%. For example, dedicating one planner 100%, from a maintenance team of 20, would increase the operating efficiency of the remaining 19 from 60% to 75%. Also, it would raise the overall weekly wrench turning hours from $20 \times 0.60 \times 40 = 480$ to $19 \times 0.75 \times 40 = 570$, for an increase of 18.75%.

1. *Time spent on planned and scheduled tasks as a percentage of total time* measures the effectiveness of the organization and maintenance planning activities. This places the focus on the work-planning phase, as planned work is typically

up to 10 times as effective as breakdown response. The planning and scheduling index measures the time spent on planned and scheduled tasks as a percentage of total work time. Notice that emphasis is placed on planning *and* scheduling. A job is planned when all the job components are worked out—what is to be done, who is to do it, what materials and equipment, and so on. Scheduling places all of these into a time slot, so that all are available when required. By themselves, planning and scheduling have a positive impact. This impact is greatly multiplied when they are combined.

Planning and scheduling index
$$= \frac{\text{Time planned and scheduled}}{\text{Total time}}$$

Time planned and scheduled = 26,000 hours

Total time = 32 employees $\times$ 2088 hours each
$$= 66,816 \text{ hours}$$

Planning and scheduling index $= \dfrac{26,000}{66,816} = 0.389$

2. *Breakdown time* measures the amount of time spent on breakdowns and, through this, indicates whether more time is needed to prevent them. Use this index in combination with other indices, because the numbers will improve as the organization becomes quicker at fixing breakdowns. This can lead to a different culture, one that prides itself on fast recovery rather than initial prevention.

Breakdown time
$$= \text{Time spent on breakdowns as \% of total time}$$

Total time = 32 employees $\times$ 2088 hours each
$$= 66,816 \text{ hours}$$

Breakdown time = 2200 hours

Breakdown time $= \dfrac{2200}{66,816} = 3.3\%$

3. *Cost of lost production due to breakdowns* is measured because the acid test of the breakdown is how much production capacity was lost. The amount of time spent by Maintenance alone shows only the time charged to the job through the Work Order, not necessarily how serious the breakdowns are. Wait time—for cooling off, restart, and materials, for example—is not included, but still prevents the equipment from operating. This measure, then, includes run-up time and the costs of lost production due to breakdown and fixing breakdowns.

Breakdown production loss
    = Cost of breakdowns as % of total direct cost

Cost of production lost time = $5140 per hour

Maintenance cost of breakdowns = $135/hr

Total direct cost = $45M

Breakdown time = 232 hours

$$\text{Breakdown production loss} = \frac{232(5140 + 135)}{45m} = 2.72\%$$

4. *The number of emergency work orders* is an overall indication of how well the breakdown problem is being kept under control. To be effective, each "emergency" must be well defined and consistently applied. (In one notable example, a realistic but cynical planner defined work priorities according to how high on the command chain the requestor was.) We advise noting measures of e-Work Order numbers in relation to a previous period (last month or last year, for example) and plot them on a graph.

Number of e-Work Orders this year = 130
Number of e-Work Orders last year = 152

3.6.4 Efficiency of Maintenance Work

This set of measures tracks the ability of Maintenance to keep up with its workload. It measures three major elements: the number of completions versus new requests, the size of the backlog, and the average response times for a request. Once again, how the measurements are made will greatly affect the results—look for consistency in the data sources and how they are measured.

1. *Work Order completions versus new requests* gives a turnover index. In the following example, an index of 107% shows that more work was completed than demanded. As a result, the backlog shrank, so overall service to customers rose. However, no account is taken of the size and complexity of the work requests or Work Orders.

WO turnover
   = Number of tasks completed as percent of work requests
   Work Orders completed last month = 3200
   Work requests last month = 3000

$$\text{WO turnover index} = \frac{3200}{3000} = 107\%$$

2. *Work Order backlog* shows the relationship between overdue Work Orders and ones completed. As with most measures, you must clearly formulate and communicate definitions. Maintenance customers will be greatly frustrated if they don't understand how this measure is created. For example, is a Work Order "overdue" when it passes the requestor's due date or the planner's? If it is the requestor's, how realistic is it?

The example shows 6.8 days—one to two weeks is generally targeted. If more than a week passes, users will complain that service is lacking. What's needed is better planning and more staff, or better screening of work requests. Less than a week suggests overstaffing.

   WOs overdue = 720
   WOs completed this month = 3200

$$\text{Backlog} = \frac{720}{3200} = 0.225 \text{ of 1 month} = 6.8 \text{ days}$$

3. *Job timeliness and response times* are measured by the time it takes from when the request is received to when the maintenance technician arrives at the job site. Comparison to a standard is the best method here. The standard should depend on the level of service desired; distance from the dispatch center to the job site; complexity of the jobs involved; availability of manpower, equipment, and materials; and urgency of the request. There is little value in measuring response times for low-priority jobs, so the trends are usually limited to emergencies and high priorities, with the statistics kept separately. In many cases, an emergency-situation response is prompted by a "work request" that comes over the phone and the responder is dispatched by phone or pager. No actual work request will be prepared, and the Work Order will be completed only when the emergency is over.

High-priority response time standard = 4 hours
High-priority response time average last month = 3.3 hours

3.6.5 Maintenance Costs

More attention is probably paid to the maintenance cost indicators than any other set of measures. This is encouraging, because the link between maintenance and costs (profits) needs to be solid and well established. As you've seen, though, many factors affect the cost of delivering maintenance services—many of them almost completely outside the control of the Maintenance Manager. Driving down maintenance costs has become a mantra in many companies, in some cases rightly so. However, reducing costs alone won't necessarily achieve your organization's objectives. Both the company's and Maintenance's mission and objectives must be factored in. One way is to relate the maintenance costs to the overall cost of production or, where single or similar product lines are produced, to the number of units produced. For example, in interdivisional or interfirm benchmarking, maintenance costs per ton of output is a widely used figure.

Within this category, many different measures are used. The examples below are of four typical ones.

Overall maintenance costs per unit output. This measure keeps track of the overall maintenance cost relative to the cost of producing the product. In a competitive environment, this is very important, particularly if the product lines are more of a commodity than a specialized product. The process industries, for instance, use these measures extensively. You can also subdivide this into the major maintenance cost components, such as direct versus overheads, materials, manpower, equipment, and contractors.

Direct maintenance cost per unit output

$$= \frac{\text{Total direct maintenance cost}}{\text{Total production units}}$$

Direct maintenance costs last month = 285,000

Total units produced last month = 6935

$$\text{DMC} = \frac{285,000}{6935} = \$41/\text{unit}$$

Stores turnover measures how effectively you use the inventory to support maintenance. *Stores value* measures the amount of materials retained in stores to service the maintenance work that needs to be done. Company policies and practices will directly affect these numbers, and limit how much the maintenance department can improve them. Similarly, if manufacturing is far from the materials source, this will also affect inventory turns. Some organizations argue that the cost of breakdown (in financial, environmental, or publicity terms) is so high that the actual stores value needed is irrelevant. Inventory turns are best measured against industry standards or last year's results. Inventory values are best measured through time.

$$\text{Inventory turnover} = \frac{\text{Cost of issues}}{\text{Inventory value}}$$

$$\text{Inventory Value Index} = \frac{\text{Inventory value this year}}{\text{Inventory value last year}}$$

Materials issued last year = \$1,400,000

Current inventory value = \$1,800,000

Last year's inventory alue = \$2,000,000

$$\text{Inventory turnover} = \frac{1,400,000}{1,800,000} = 0.8 \text{ turns}$$

$$\text{Inventory Value Index} = \frac{1,800,000}{2,000,000} = 0.9$$

Note that an inventory index of less than 1 indicates a reduced inventory value, or an improvement in performance, while an increase in the inventory turns shows better use of the organization's inventory investment.

Maintenance cost versus the cost of the asset base. This measures how effectively the maintenance department manages to repair and maintain the overall asset base. It uses the asset or replacement value to make the calculation, depending on available data. As with many of the cost measures, it can be cut several ways to select individual cost elements. The final measure shows the percentage of the asset's value devoted to repair and maintenance.

$$\text{Direct maintenance cost-effectiveness} = \frac{\text{Total direct maintenance cost}}{\text{Asset value (or Replacement cost)}}$$

These can have quite significantly different values, e.g., book value compared to replacement cost.

Direct maintenance cost = \$4,000,000

Asset value = \$40,000,000

$$\text{DMCE} = \frac{4,000,000}{40,000,000} = 10\%$$

Overall maintenance effectiveness index shows the rate at which Maintenance's overall effectiveness is improving or deteriorating. It compares, from period to period, costs of maintenance plus lost production. The overall index should go down, but increased maintenance costs or production losses due to maintenance will force the index up. If a new maintenance program is introduced, this index will measure whether spending more on maintenance has paid off in lower production losses.

Maintenance Improvement Index
$$= \frac{\text{Total direct maintenance cost} + \text{Production losses previous month}}{\text{Total direct maintenance cost} + \text{Production losses last month}}$$

$$\text{Maintenance Improvement Index} = \frac{285,000 + 102,800}{270,000 + 128,500} = 0.97$$

Note that results below 1.0 show a deterioration in the index.

3.6.6 Maintenance Quality

The pundits frequently ignore maintenance quality when looking at performance measures. Most of the auto magazines, however, feature this in their "which car to buy?" columns. Use it to judge how often repeat problems occur and how often the dealer can fix the problem on the first visit.

The maintenance department can collect and measure this data in several ways. At least one CMMS has a special built-in feedback form that is sent automatically to the requestor when the work is completed.

Repeat jobs and repeat breakdowns generally indicate that problems haven't been correctly diagnosed or training and/or materials aren't up to standard. Many maintenance departments argue that most repeats occur because they can't schedule the equipment down for adequate maintenance. The most effective way to get enough maintenance time is to measure and demonstrate the cost of breakdowns. Note that *repeat* refers only to corrective and breakdown work, not to preventive or predictive tasks.

$$\text{Repeat Jobs Index} = \frac{\text{Number of repeat jobs this year}}{\text{Number of repeat jobs last year}}$$

Number of repeat jobs this year = 67

Number of repeat jobs last year = 80

$$\text{Repeat Jobs Index} = \frac{67}{80} = 0.84$$

Stock-outs constitute one of the most contentious areas, reflecting tension between Finance—which wants to minimize inventory—and Operations—which, to maintain output, needs spares to support it. You can maintain the balance with good planning—predicting when materials will be needed, knowing the delivery times, and adding a safety margin based on historical predictions. Stock-outs are normally measured against the previous period, but are best when tied to the higher-priority work. Zero stock-outs isn't necessarily good; it may indicate overstocking.

$$\text{Stock-out Index} = \frac{\text{Stock-outs this year}}{\text{Stock-outs last year}}$$

Stock-outs this year = 16

Stock-outs last year = 20

$$\text{Stock-out Index} = \frac{16}{20} = 0.8$$

Work Order accuracy measures how closely the planning process from work request to job completion matches the reality. The core of the process is applying manpower and materials to jobs, and this is usually the focus of this measure. You can easily measure this using the comment section on the Work Order. Encourage the maintenance technician to provide feedback on job-specification errors, skill requirements, and specified materials so corrections can be made for the next time around.

Work Order accuracy
$$= 1 - \frac{\text{Number of Work Orders completed}}{\text{Number of Work Order errors identified}}$$
Number of Work Orders completed = 1300

Number of Work Order errors identified = 15

Work Order accuracy $= 1 - \dfrac{15}{1300} = 98.85\%$

3.7 COLLECTING THE DATA

From the above examples, you can select measures that will generate the right information to drive action. The data is drawn from condition-based monitoring systems, Enterprise Asset Management systems, engineering systems, and process-control systems. You don't need all these systems, however, to start the performance-evaluation process. Most of the data is also available from other sources, although computerized systems make data collection much easier. Table 3-5 shows a core data set for all of the examples in Section 3.6.

The volume of available data is expanding rapidly. Historically, the Maintenance Manager's problem was not having enough data to make an informed decision. With today's various computerized systems, the reverse is the problem: too much data. One possible solution to this is the Maintenance Knowledge Base being developed by at least one EAM company and one CBM company. This concept recognizes that not only must various data sources be identified; more than a simple point-to-point linkage between these sources is needed (Figure 3-14). To achieve full value from the data, a knowledge base must be constructed to selectively cull the data, analyze

Table 3-5   Core Data Set for Performance Indicators

Item	Value
Total time—full-time operation	$7 \times 24 \times 365 = 8760$ hours/year
Downtime	
Scheduled	160 hours/year
Unscheduled	232 hours/year
Nonutilized time	600 hours/year
Number of scheduled outages	6/year
Number of breakdowns	12/year
Production rate (units per hour)	10/hour
Annual production (units)	
Capacity	87,600 units
Actual	77,680 units
Units produced last month	6935
Number of maintenance people	32
Working hours per person per year	2088 hours
Maintenance hours—capacity	66,816 hours
Total wrench time	46,100 hours
Planned time	44,700 hours
Time scheduled	26,000 hours
Total maintenance costs	$4.0 M
Total direct maintenance costs/year	$3.2 M
Total direct maintenance costs	
Last month	$270,000
Previous month	$285,000
Total materials issued per year	$1.4 M
Total manpower costs per year	$1.6 M
Total Work Orders per year	32,000
Total work requests last month	3000
Work Orders completed last month	3200
Breakdown hours worked	2200
Work Order errors	15
Repeat jobs	
This year	67
Last year	80
Overdue work orders	720
Emergency Work Orders	
Last year	152
This year	130

Table 3-5   Continued

Item	Value
High-priority response time	
Standard	4 hours
Last month	3.3 hours
Maintenance cost of breakdowns	$135/hour
Stores value	
Last year	$1.8 M
Current	$2.0 M
Stock-outs	
This year	16
Last year	20
Total production costs	$45 M
Lost production time cost/hour	$5140
Production losses	
Last month	20 hours
Previous month	25 hours
Asset value (replacement cost)	$40 M

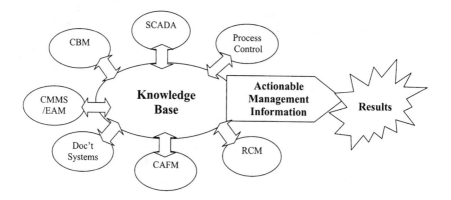

Figure 3-14   The knowledge base.

it, and use it as a decision-support tool. That is where the real value lies—helping to develop actionable management information that attains results. Without this, the data isn't particularly useful.

3.8 APPLYING PERFORMANCE MEASUREMENT TO INDIVIDUAL EQUIPMENT

So far, all the performance indicators we have reviewed apply to the broader spectrum of maintenance costs and operations. There are also many indicators that effectively apply at the micro level to individual equipment or jobs. Particularly if the organization has numerous examples of the same piece of equipment running, comparative evaluations can show the varied operating results and costs caused by different running conditions and maintenance methodologies. Adopting an internal Best Practices approach in your organization can yield significant improvements.

To do this type of forensic maintenance, the detailed data must be readily available, and almost demands you use an EAM or CBM system. The data needs to be accessed in large enough sample sizes to reduce the error probability to acceptable levels. For this, the data typically has to be aggregated from individual Work Orders, pick lists, condition reports, and process-control data sheets. Manual collecting from these sources isn't really feasible. Two examples will be enough.

Looking at maintenance analysis at this level, you're typically seeking specific results, relating to optimal maintenance intervals, operating parameters, and cost savings. Figure 3-15 tracks the energy savings from compressors as various simple maintenance tasks were done. The improvement in energy costs (i.e., energy cost savings) totaled an annual $6000, representing a savings of about 7.5% of the total annual "before" cost of $81,000.

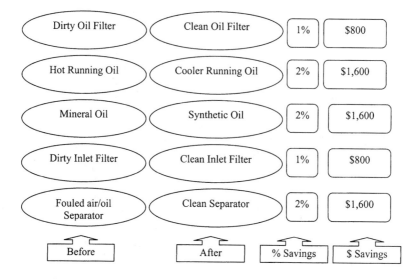

Figure 3-15 Energy cost savings generated from maintenance tasks.

The following example uses life-cycle costing involving a replace-versus-repair decision. With the repair alternative, a slurry pump can continue to operate for five more years, but with a higher annual maintenance cost and likelihood of breakdown. To offset this, there will be no purchase or installation costs (Table 3-6).

The case for a replacement appears to be clear-cut, but you need to ask the same qualifying questions as before to be sure you make the "correct" decision. These relate to capacity, customer satisfaction, and failure-data reliability. Also, ask additional questions such as:

- The Repair case covers a five-year planning period, versus 12 years in the Replace case—is this significant?
- What is the decision-making impact of the zero purchase cost in the Repair case—should a capital cost be included?

Table 3-6   Replace Versus Repair Decision Analysis

	Repair	Replace
Purchase cost	0	$20,000
Cost to install	0	$2000
Annual running cost	$3000	$4000
Annual maintenance cost	$9000	$8000
Final disposal cost less scrap value	$1000	$500
Pump life (years)	5 years	12 years
Total life-cycle cost	$61,000	$167,000
Annual cost	**$12,200**	**$13,900**
Average throughput (gals/hr)	100	175
Lifetime throughput (M gals)	4.38	18.40
Cost (cents/gal)	**1.39**	**0.91**
Average breakdown frequency	3 per year	2 per year
Average breakdown duration	1 day	1 day
Downtime cost ($/hour)	$1000/hour	$1000/hour
Downtime cost (total lifetime)	$360,000	$576,000
Downtime cost (cents/gallon)	8.22	3.13
Total operating costs (cents/gallon)	**9.61**	**4.04**

- Does the fact that the funds come from two budgets affect the decision—i.e., purchase price from the capital budget, running and maintenance costs from the operating budget?

To do any kind of meaningful analysis, the base data must be readily available. Critics frequently and successfully challenge the results based on the integrity of the data. When setting up your data-collection process, make it easy to record, reliable, and consistently analyzed.

The other key issue is where to start. Every Maintenance Manager needs more time and less work, but you know that making time only comes from doing things more effectively. Figure 3-16 will help you sort out where to start. Grade the measures and actions high to low, based on how much benefit they create and how difficult they are to implement. Then start with those in the top left quadrant: high payoff, low implementation difficulty.

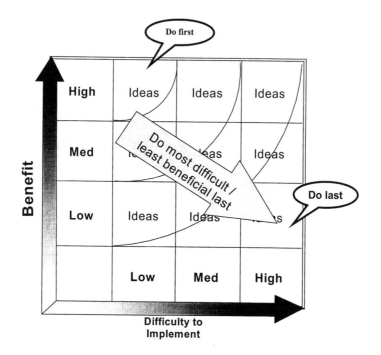

Figure 3-16   Benefits and implementation difficulty matrix.

3.9   WHO'S LISTENING? TURNING
MEASUREMENTS INTO INFORMATION

A frequent complaint from the maintenance department is that "they" won't listen. Who are *they*? Is it Management deciding not to release funds for needed improvements? Is it Engineering refusing to accept the maintenance specifications for the new equipment? Is it Finance insisting on return on investment calculations at every turn? Is it Purchasing continuing to buy the same old junk that we know will fail in three months? Or is it Production, who won't give adequate downtime for maintenance?

In fact, it is all of them, and the core reason has usually been the same: Maintenance hasn't been able to make its case convincingly. For that, you need facts, figures, and attractive graphs. The cynics claim that graphs were invented because Management can't read. There is more than a grain of truth in this, although it stems from lack of priority, not of time or ability. The trick is to make the results both attractive and compelling enough to make them a priority and involve the other departments in the process. Buy-in in your organization can work both upward and sideways, as well as downward.

For example, corporate finance has become much more complex and sophisticated than it used to be. Many of the issues that make Finance question the maintenance department's proposals are technical, requiring input from Accounting. Co-opt a finance person onto the team to handle skill-testing questions such as:

- How are inflation and the cost of money handled?
- What are the threshold project Return on Investment (ROI) levels?
- How is ROI calculated?
- What discount rate is used?
- How far out should you project?
- What backup for the revenue, savings, and cost figures is needed?
- How do you deal with the built-in need for conservatism?
- How do you accommodate risk in the project?

A parallel set of questions will arise with the engineering, systems, and other groups. Involve them in the process when their help is needed to clear barriers.

3.10 EIGHT STEPS TO IMPLEMENTING A MEASUREMENT SYSTEM

The performance-measurement implementation process follows a straightforward and logical pattern. However, because

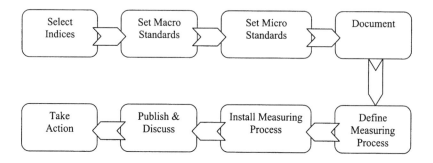

Figure 3-17   Steps to implement a measurement system.

many organizations don't do a full implementation, they get a methodology that can't meet the daily demands of the average maintenance department. The major steps are highlighted in Figure 3-17.

1.   Review and select the indices that make sense for your organization. As shown earlier in this chapter, there are literally hundreds of measurements and performance indicators that can be used. The starting point is to understand the structure and purpose of the more commonly used ones. Typically, you undertake a series of in-house round tables and workshops to initiate discussion among maintenance personnel.

Once the basics are reasonably well understood, choose a small cross-section of measurements. Others can be added later. Select the chosen few based on their fit with the department's objectives, measurability, and ability to affect the results. At the early stages, keep to a minimum the amount of extra work to build the performance measures. As the payoffs start to appear, the effort can be expanded. Also, make sure that the data for the early measurements is readily available, and that the source data is accurate.

2.   For each of the chosen measurements, set standards, values, and targets. To ensure reasonably full coverage, choose at least one from each of the six macro categories:

- Maintenance results
- Maintenance productivity
- Maintenance organization
- Efficiency of maintenance work
- Maintenance costs
- Maintenance quality

You will have to do some research to come up with valid targets that relate to your organization's current numbers, plus the hoped-for future improvements. Beware of setting targets that are impossible to reach.

3. Supplement the macro measures with a small number of measures for critical equipment and bad actors at the micro level. Set standards and targets for each—again, paying attention to the data availability and consistency.

4. Document the measures and targets, plus the interpretation of each number or trend. Absolute numbers may not mean anything; so whenever you use a trend or index, explain it (i.e., is the level good or bad?).

5. Setting up the measuring process, make sure that the person responsible clearly understands the nature of the measurement, where the data comes from, and what sort of analysis is required. Include how often the reading is to be taken and, if relevant, whether it should be at a specific time or event during the day (end of batch, for example). Consistency needs to be stressed. Erratic results not only will sully the measurement but may induce the wrong action.

With the widespread use of CMMS, EAM, and CBM systems, the amount of available data has grown dramatically. System capability to collect and maintain the right data, however, varies just as dramatically. Once you've established the data set, make sure the computer system makes it easy to accurately collect the data. The Work Order data fields, for example, should carry the same labels as the measurement process. They should be mandatory fields so the maintenance technician can understand why they are needed. Build in a value range so that any data entry outside of it will immediately be flagged.

6. Install the measuring procedure in the CMMS or EAM system as a regular weekly Work Order. The frequent omission of this simple step is a major reason why measurement systems fail.

7. Publish the results so that all maintenance employees and visitors can see both the targets and the achievements. This is best done on the departmental notice boards, although some maintenance departments have set up a special war room to display and discuss the results. Set aside time at the weekly meetings to review and discuss the results, especially looking for new ideas to achieve the results.

8. The point of taking measurements is to target trend implications and remedial actions to make improvements. A milestone flag on the trend chart is an effective way to show when a specific action was taken, and the subsequent impact. As with implementation of the measurement program, the remedial action should emphasize what has to be done, who does it and when, and what materials and tools are needed. In fact, what's needed is a regular Work Order to record the task for future evaluation.

Most often, measurement projects are seen as just that— measurement projects. But a measurement project can be much more than that. It can be a dynamic program for ongoing change. To make this happen, use the measurement results as the basis to re-evaluate the macro and micro standards and targets. Establish a PM-type Work Order calling for an annual review of each measurement and set new targets to introduce this feedback loop.

3.11 TURNING MEASUREMENTS INTO ACTION

Measurements are only as good as the actions they generate. If you fail to convert them into action, you miss the whole point of performance measurement. The flip side is the tendency for a good trend to lead to complacency. Some organizations have rejected the basic philosophy of benchmarking, for example, claiming that it forces the organization into a perpetual catch-up mentality. No company ever took the lead playing catch-up. The breakaway firm needs to think outside the box, and the measurement process helps to define the size and shape of the box.

The final word on this phase of the measurement issue goes to Terry Wireman:

> Yesterday's excellence is today's standard and tomorrow's mediocrity.

3.12 ROLE OF KEY PERFORMANCE INDICATORS: PROS AND CONS

Management uses performance measurement primarily for monitoring purposes, and many performance indicators have been developed to support operational decisions. These indicators are, at best, descriptive signals that some action needs to be taken. To make them more useful, put in place decision rules that are compatible with organizational objectives. This way, you can determine your preferred course of action based on the indicators' values.

To clarify trends when the activity level may vary over time, or when comparing organizations of different size, you can use indices to measure maintenance performance. Campbell (4) classifies these commonly used performance measures into three categories, based on their focus:

1. Measures of equipment performance—e.g., availability, reliability, and overall equipment effectiveness
2. Measures of cost performance—e.g., operation and maintenance (O&M) labor and material costs

3. Measures of process performance—e.g., ratio of planned and unplanned work, schedule compliance

However, the underlying assumptions of these measures are often not considered when interpreting results, so their value can be questionable. For example, traditional financial measures still tend to encourage managers to focus on short-term results, a definite drawback. This flawed thinking is driven by the investment community's fixation with share prices, driven largely by the current quarter's earnings. As a result, very few managers choose to make (or will receive Board approval for) capital investments and long-term strategic objectives that may jeopardize quarterly earnings targets.

Income-based financial figures are lag indicators. They are better at measuring the consequences of yesterday's decisions than indicating tomorrow's performance. Many managers are forced to play this short-term earnings game. For instance, maintenance investment can be cut back to boost the quarterly earnings. The detrimental effect of the cutback will show up only as increased operating cost in the future. By then, the manager making the cutback decision may have already been promoted because of the excellent earnings performance. To make up for these deficiencies, customer-oriented measures such as response time, service commitments, and satisfaction have become important lead indicators of business success.

To assure future rewards, your organization must be both financially sound and customer-oriented. This is possible only with distinctive core competencies that will enable you to achieve your business objectives. Furthermore, you must improve and create value continuously, through developing your most precious assets: your employees. An organization that excels in only some of these dimensions will be, at best, a mediocre performer. Operational improvements such as faster response, better quality of service, and reduced waste won't lead to better financial performance unless the spare capacity they create is used or the operation is downsized. Also, maintenance organizations that deliver high-quality services won't remain viable for long if they are slow in developing expertise to meet the emerging needs of the user departments. For example, electromechanical systems are being phased out by electronic and software systems in many automatic facilities. In the face of the new demand, the maintenance service provider has to transform its expertise from primarily electrical and mechanical trades to electronics and information technology.

Obviously, you won't fulfill all these requirements by relying on a few measures that represent a narrow perspective. You need a balanced results presentation to measure maintenance performance. The Balanced Scorecard (BSC) proposed by Kaplan and Norton (5) offers the template for the balanced presentation. The BSC translates a business unit's mission and strategy into objectives and quantifiable measures built around four perspectives:

- Financial (the investor's views)
- Customer (the performance attributes valued by customers)
- Internal processes (the long- and short-term means to achieve the financial and customer objectives)
- Learning and growth (capability to improve and create value)—it focuses managers on a handful of critical measures for the organization's continued success

The Balanced Scorecard has been implemented in numerous major engineering, construction, microelectronics, and computer companies. Their experience indicates that the BSC's greatest impact on business performance is to drive change process. The BSC promotes a strategic management system that links long-term strategic objectives to short-term actions (5).

A strategic management system built around a BSC is characterized by three keywords: focus, balance, and integration. Ashton (6) explains these three attributes:

Focus has both strategic and operational dimensions in defining direction, capability and what the business or its activities are all about, while *balance* seeks an equilibrium for making sense of the business and to strengthen focus. *Integration* is critical, ensuring that organizational effort knits into some form of sustainable response to strategic priority and change.

The BSC is a complete framework for establishing performance-management systems at the corporate or business-unit level. When the approach is applied to managing maintenance performance, follow the process outlined below (7).

1. Formulate maintenance operation strategy. Consider strategic options such as developing in-house capability, outsourcing maintenance, empowering front-line operators to practice autonomous maintenance, developing a multiskilled maintenance workforce, and implementing condition-based maintenance. Get others involved in making decisions.

2. Operationalize the strategy. Translate the maintenance strategy into long-term objectives. Identify the relevant Key Performance Indicators (KPIs) to be included in the BSC and establish performance targets. Suppose an electric utility company has chosen to outsource its maintenance and repair of generic and common equipment and vehicle fleets so it can manage its core transmission and distribution system. The KPIs and performance targets that relate to this strategic objective are to outsource 20% of maintenance work and reduce maintenance costs by 30% in 2 years. The former indicator belongs to the Internal Processes perspective and the latter to the Financial perspective. To achieve vertical alignment, these objectives, KPIs, and targets are aligned into team and individual goals.

3. Develop action plans. These are means to the ends stipulated in the targets established in Step 2. To reach the outsourcing targets in the above example, your company may develop capabilities in the following outsourcing areas: contract negotiation, contract management, and capitalizing on emerging technology and competitive opportunities in maintenance. These action plans should also include any necessary changes in your organization's support infrastructure, such as maintenance work structuring, management information systems, reward and recognition, and resource-allocation mechanisms.

4. Periodically review performance and strategy. Track progress in meeting strategic objectives and validate the causal relationships between measures at defined intervals. After the review, you may need to draft new strategic objectives, modify action plans, and revise the scorecard.

Some of the BSC KPIs that measure the maintenance performance of an electricity transmission and distribution company may include those shown in Table 3-7 (8). Since these measures reflect an organization's strategic objectives, the BSC is specific to the organization for which it is developed.

By directing managers to consider all the important measures together, the BSC guards against suboptimization. Unlike conventional measures, which are control-oriented, the BSC puts strategy and vision at the center and emphasizes achieving performance targets. The measures are designed to pull people toward the overall vision. To identify them and establish their stretch targets, you need to consult with internal and external stakeholders—senior management, key personnel in maintenance operations, and maintenance users. This way, the performance measures for the maintenance operation are linked to the business success of the whole organization.

Table 3-7   Linking Strategic Objectives to Key Performance
Indicators

Perspective	Strategic objectives	Key Performance Indicators (KPIs)
Financial	Reduce operation and maintenance (O&M) costs	O&M costs per customer
Customer	Increase customer satisfaction	Customer-minute loss; customer satisfaction rating
Internal processes	Enhance system integrity	% of time voltage exceeds limits; number of contingency plans reviewed
Learning and growth	Develop a multiskilled and empowered workforce	% of cross-trained staff; hours of training per employee

Source: Ref. 8.

The theoretical underpinning of the BSC approach to measuring performance is built on two assertions:

1. Strategic planning has a strong and positive effect on a firm's performance.
2. Group goals influence group performance.

The link between strategic planning and company performance has been the subject of numerous research studies. Miller and Cardinal (9) have applied the meta-analytic technique to analyze empirical data from planning performance studies of the last two decades, establishing a strong and positive connection between strategic planning and growth. They also show that, when a company is operating under turbulent conditions, there is a similar link between planning and profitability. A similar study on previously published research findings (10) confirms the group goal effect.

Although industry commonly agrees that strategic planning is essential for future success, performance measures and actual company improvement programs are often inconsistent with the declared strategy. This discrepancy between strategic intent and operational objectives and measures is described in a Belgian manufacturing survey (11). You can ensure that you don't make the same mistake by introducing the Balanced Scorecard.

REFERENCES

1. Jasper L. Coetzee. Maintenance. Maintenance Publishers, 1997.
2. Terry Wireman. Developing Performance Indicators for Managing Maintenance. New York: Industrial Press, 1998.
3. MIMOSA: Machinery Information Management Open Systems Alliance, www.mimosa.org
4. John D. Campbell. Uptime: Strategies for Excellence in Maintenance Management. Portland, OR: Productivity Press, 1994.
5. R.S. Kaplan, D.P. Norton. The Balanced Scorecard. Boston, MA: Harvard Business School Press, 1996.
6. C. Ashton. Strategic Performance Measurement. London: Business Intelligence Ltd., 1997.
7. A.H.C. Tsang. A strategic approach to managing maintenance performance. Journal of Quality in Maintenance Engineering 4(2):87–94, 1998.
8. A.H.C. Tsang, W.L. Brown. Managing the maintenance performance of an electric utility through the use of balanced scorecards. Proceedings of 3rd International Conference of Maintenance Societies, Adelaide, 1998, pp. 1–10.
9. C.C. Miller, L.B. Cardinal. Strategic planning and firm performance: a synthesis of more than two decades of research. Academy of Management Journal 37(6):1649–1665, 1994.
10. A.M. O'Leary-Kelly, J.J. Martocchio, D.D. Frink. A review of the influence of group goals on group performance. Academy of Management Journal 37(5):1285–1301, 1994.
11. L. Gelders, P. Mannaerts, J. Maes. Manufacturing strategy, performance indicators and improvement programmes. International Journal of Production Research 32(4):797–805, 1994.

4

Data Acquisition

We live in an information age. Today, more than ever, decisions are driven by hard information, derived from data. In asset management, particularly maintenance, you need readily available data for the kind of thorough analysis that produces optimal solutions. This chapter discusses the key aspects of modern computer-based maintenance management systems:

- Maintenance Management Systems—CMMS, EAM
- Evolution and direction of Maintenance Management Systems
- System selection criteria
- System implementation

First, we explore Computerized Maintenance Management Systems (CMMS) and Enterprise Asset Management (EAM) systems—how they work and how they can help you attain maximum results. Often, the terms CMMS and EAM system are used interchangeably. But you will see that a CMMS, the packaged software application, is really the enabler for EAM, just as a word processor enables you to write on a computer.

Next, we examine how maintenance management systems have evolved to their current state, both functionally and technically. These systems continue to develop and, as we explain, their potential benefits are increasing.

Once you decide to acquire a Maintenance Management System, you want to ensure that it will produce tangible business improvements. We look at the two crucial areas you must get right—selecting and implementing the system—to achieve the results you need.

4.1 DEFINING MAINTENANCE MANAGEMENT SYSTEMS

First, what *is a* maintenance management system? Nearly everybody is familiar with an accounting system, and there isn't a CFO or controller who doesn't daily use accounting data. Investment decisions are made with a keen eye on the balance sheet, income statement, and, of course, shareholder perception. The accounting system, in effect, monitors the health of the business, and there are dozens of standard measures to work with.

At an individual level, banks have moved toward "personal" financial systems that ideally record and categorize every expenditure, investment, and income stream. They provide software that enables Internet transactions so that you can now better plan expenditures, budget for vacations, and track savings. The growth of computer based personal financial systems has resulted in, and encouraged, greater numeracy.

In the same way, a Maintenance Management System enables you to monitor the asset base, the value of which, especially for resource development and manufacturing enterprises, can add up to billions of dollars.

As noted above, Maintenance Management Systems are often referred to as Computerized Maintenance Management Systems (CMMS) and Enterprise Asset Management (EAM) systems. The difference is essentially one of scale, and there is no clear dividing line. The choice usually comes down to an organization's asset-management philosophy and strategies. To help decide, look at Maintenance Management Systems as an enabler for total enterprise asset management. If your enterprise spans many plants, and several jurisdictions and nationalities, you need sophisticated, high-end CMMS applications. At this level, only CMMS can seamlessly integrate with other business applications (financial, human resources, procurement, security, material planning, capital project man-

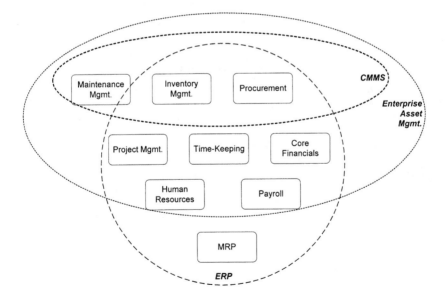

Figure 4-1 Comparison between CMMS, EAM, and ERP.

agement, etc.) to produce a total business information solution.

Figure 4-1 clarifies the similarities and differences between CMMS and EAM system. When it comes to maintenance management, some systems, although properly called Enterprise Resource Planning (ERP), do not include this function, so the component is shown partially enclosed.

4.2 EVOLUTION OF MAINTENANCE MANAGEMENT SYSTEMS

Maintenance Management Systems have developed, over a long time, out of focused business needs. Table 4-1 shows how key functions have evolved. The earliest maintenance management applications were custom built to specification. From these pioneer systems, often well built but costly, descend most current packaged applications. There were huge differences between these initial systems in their functions and implementation, just as in the first PC applications. At first,

Table 4-1   Functional Developments and Trends in CMMS

From	To
Custom-built	Package solution
Clear leaders	Many common features
Difficult to interface with other business applications	Integration with ERP systems
Single "site"	Enterprise-wide
Narrow focus (work management)	Total asset-management functionality
Difficult to modify	Easily customized and configured
	Deployed with embedded/integrated Predictive Maintenance (PdM), Condition-Based Monitoring (CBM), and Reliability-Centered Maintenance (RCM) functions

there were clear leaders but, over time, the best features of each product became standard. Vendors consolidated and the selection process became more complex. Today you need to do a detailed comparison to detect the differences.

The earlier systems were built to suit each customer's specific needs, both functional and technical. As a result, they couldn't easily be deployed elsewhere. This changed with the advent of the many de facto standards—MS-DOS, Windows, relational databases, Structured Query Language (SQL), etc. Today, it's mandatory to share data with other business applications for all but the smallest CMMS applications.

As organizations grow and evolve over time, so should their core business systems. Consolidation, particularly, has driven the need for business systems that are true enterprise applications, dealing with multiple physical plants, sites, currencies, time zones, and even languages. World-class ERP systems offer this kind of location transparency and the leading maintenance management applications are now comparably complex.

The scope of leading maintenance management applications is very wide, incorporating modules for:

- Maintenance (work order management, scheduling, estimating, workflow, preventive maintenance, equipment hierarchies, equipment tracking, capital project management)
- Inventory management (parts lists, repairable items management, catalogs, warehouse management)
- Procurement (PO processes, vendor agreements, contractor management and administration)
- Human resources (health and safety, time control, skills management, payroll, benefits, recruitment, training)
- Financials (general ledger accounts, payable accounts, receivable, fixed assets, activity-based costing, budgeting)

Also, the leading maintenance management vendors offer sophisticated performance measurement and reporting.

This wide range is in contrast to earlier maintenance management applications, which focused solely on tactical work management—usually work order initiation, resources required, processing, and closing.

Finally, modern asset management is turning to optimization methods to improve maintenance effectiveness. Techniques such as condition-based monitoring (where maintenance is condition-driven rather than time-interval-driven), Reliability-Centered Maintenance (described in Chapter 7), and optimal repair/replace decisions draw on rich stores of historical information in the Maintenance Management System database. Using these techniques requires additional software modules, now being embedded in the leading maintenance management products.

Technically, there have been several significant transitions, as shown in Table 4-2, in which you can see a gradual standardization process developing. Certainly, adopting the relational database "standard" has made intersystem communication much easier. As well as a new architecture for distributed systems, the Internet provided Transmission Connection Protocol/Internet Protocol (TCP/IP), the current communication standard for networks. This, in turn, has allowed systems to interconnect as never before.

Table 4-2   Technical Developments in CMMS
Applications

From	To
Mainframe	Micro/Mini
Data files	Relational database
Terminal/file server	Client/server
Proprietary	Open standards
Dedicated infrastructure	TCP/IP/Internet enabled
Paper help	Context-sensitive/on-line help
Classroom training	Computer-based training

It used to be that racks of paper information for large business systems would sit in the system administrator's office, virtually inaccessible to users. This has entirely disappeared and current applications now have extensive on-line help systems. Nevertheless, as application complexity increases, training continues to be a problem.

You can't consider current business systems without including the Internet and electronic-enabled initiatives (in this chapter, the word Internet is used for all related network terms such as intranet and extranet). Today, it would be difficult to find a maintenance management system Request for Proposal (RFP) that didn't specify the need for Web-enabled functionality. But one of the greatest difficulties facing Maintenance Management System specifiers is understanding what that really means and the associated tradeoffs. There are many opportunities for Web-enabling a Maintenance Management System, notably:

- Complete application delivery, including application leasing
- Internet-enabled workflow (e-mail driven)
- Management reporting
- Supplier management/procurement

A brief description of each initiative follows.

Application delivery. As long as your hardware infrastructure is sufficiently robust (bandwidth, reliability, performance, etc.), you can deploy some CMMS products completely through a standard Web browser. The advantages of this include version control, potentially smaller (less powerful, therefore cheaper) desktop machines, and ease of expansion. A recent twist on this is to lease, not buy, the CMMS application. This is akin to mainframe "time-sharing," which is still used for payroll processing and similar functions.

Internet-enabled workflow. One of the most popular Maintenance Management System features is comprehensive workflow support, so that business rules can be written into the system and changed at will. By including Internet technology, you can expand the scope of your business rules to cover the globe and, if you can imagine, perhaps even farther. There is ongoing research into expanding the Internet to include spacecraft and extraterrestrial sites!

Management reporting. Closer to home, the Internet is ideal for disseminating management reports and, in fact, is a growing part of true Enterprise Asset Management. You need current data to make better business decisions. In some cases, data even a week old is out of date. Being able to access an up-to the-minute corporate database from any Internet connection is becoming a competitive necessity.

Supplier management and e-procurement. This has received a lot of attention, primarily because it is so Internet-driven. Whether you're buying capital items or office supplies, the basic procurement cycle is essentially the same. E-procurement automates time- and labor-intensive processes, as well as enforcing a single enterprise-wide policy through a single buying interface.

E-procurement includes: on-line supplier catalogs (several new companies have sprung up to provide them), requisitions sent over the Internet, POs, receipt and billing confirmation (essentially Electronic Data Interchange, or EDI, with a new twist), and guaranteed security for financial transactions. We are only skimming the surface here—entire books have been written on these topics.

4.3   ASSET MANAGEMENT
SYSTEM SELECTION

If your organization decides to acquire a new Asset Management System, and to make you responsible for selecting and deploying it, this could be the career opportunity of a lifetime. However, you need to ask some key questions before you proceed further. Once you receive satisfactory answers, the next requirement is a robust system selection methodology roadmap.

4.3.1 Preliminary Considerations

For the most part, you use computer-based business solutions to increase your organization's effectiveness. Put simply, the solution has to enhance profitability. Normally, you make a business case to describe, justify, and financially estimate the expected benefits. As shown in Figure 4-2, acquiring a business system like a CMMS is usually part of an improvement cycle. In this cycle, the business case is part of the remediation plan, along with other future success measures.

For example, management may expect the CMMS to improve productivity and Overall Equipment Effectiveness (OEE) by a certain percent, increase inventory turns, and so on. The team responsible for selecting the CMMS needs to understand not only these specific goals but the overall strategic context for acquiring the CMMS.

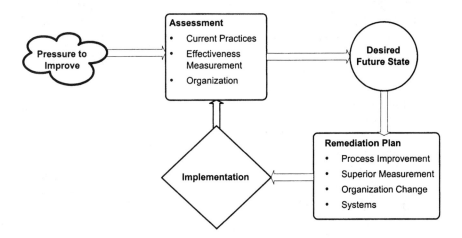

Figure 4-2 Business improvement cycle.

4.3.2 System Selection Process

Most packaged systems are selected in a similar manner. Unlike custom systems, when you procure a packaged system you decide what you need and then evaluate various vendor offerings to see which fits best. The general approach is illustrated in Figure 4-3.

In the sections that follow, we describe each step in the process. In practice, most organizations use an outside consultant to either help them execute the process or take it over completely. The advantage is that the consultant can fast-track many tasks, minimizing the cost of disrupting the organization.

4.3.2.1 Establish Teams

Your goal at this point is to establish working teams to set requirements and validate vendor offerings against them. Teams usually include users, maintenance managers, and others who have a large stake in the success of the CMMS.

If the scope of the planned CMMS is truly enterprise-wide, you may need several teams, representing different plants, sites, or locations. If the CMMS will be complex, consider forming teams with particular domain expertise. There could be teams from maintenance, inventory management, procurement, and so on. Clearly, as the number of teams

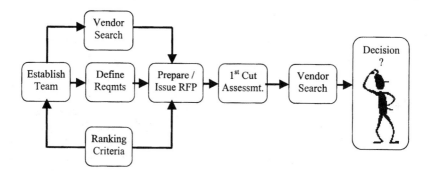

Figure 4-3   System-selection roadmap.

grows, so does the task of coordinating their outputs into cohesive requirements. You need, then, to form each team carefully, with a clear definition of what it's expected to deliver.

The key deliverables are:

- Project "charter"—what teams are expected to deliver, in how much detail. The charter should also define each team's specific responsibilities.
- Task schedule—the timeline for each team. Don't expect results after only a couple of meetings, since capturing requirements can be onerous.

The overall manager of this task should recognize that team-building skills will be needed and he or she will frequently have to adjudicate in situations where responsibilities aren't clear.

4.3.2.2 Vendor Search

Assemble an initial list of prequalified vendors you're inviting to bid. Typically, this is done with a Request for Information (RFI), which must include an outline, at least, of the CMMS scope, to get the process moving and obtain consensus. At this stage, you probably won't be able to decide based on functionality alone, since the majority of the prime systems provide more functions than even the most advanced user needs. But if you have clear functional requirements that point to certain vendors, take that into account.

An RFI should also ask about the vendors' commercial and financial viability, track records of comparable installations (particularly if the proposed installation sites are far from the vendors' home offices), product support capability, and other "due diligence" considerations.

You should issue, return, and analyze the RFI in time to meet the Request for Proposal's planned issue date. Usually, you won't need all the teams to accomplish this, probably only team leaders.

To recap, typical deliverables from this task are:

- Request for Information
- Initial vendor list

4.3.2.3   Define Requirements

The quality of vendor proposals will largely reflect how complete and clear your requirements document is. Also known as the system specification, it is the core document against which the CMMS application is acquired, implemented, and tested. Obviously, it needs to be assembled with care.

At the highest level, group requirements into major categories. Then break them down into subcategories and again, if necessary, into very specific requirement criteria. Table 4-3 is an example of what a requirements hierarchy can look like. Note that where detailed requirements are stated as questions, they should be presented in an RFP in the form "the system shall . . ." If you're uncertain about specific needs, pose the requirement as a question, inviting vendor comment.

Requirements are best gathered using business process maps as the context. You should have generated these as part of the Assessment and Desired Future State work shown in Figure 4-2. For example, consider work order processes—raising, approving, executing, closing, reporting, etc. With these maps, the team(s) responsible for requirements gathering should set up workshop interviews with user groups. A workshop format, bringing together different perspectives, stimulates maximum input.

4.3.2.4   Define Ranking Criteria

At the end of the exercise, the requirements hierarchy will be very large, with potentially hundreds of detailed system requirements across all of the major categories. You will need a quantitative approach to compare all the vendor responses.

Two numerical scores are relevant to each specification item: degree of need and degree of compliance. Degree of need represents how badly you must have the specification item, from mandatory to "would be nice." Clearly, a mandatory requirement should be scored higher than one that is optional. In the example above, the requirement for the system to support multiple warehouses would probably be mandatory and scored, say, 5 in degree of need. In contrast, if supporting a warehouse hierarchy is optional, it could be scored 3, representing a "highly desirable" requirement. An unimportant specification item would be scored at only 1. To make the rating even simpler, you can use a binary score: 5 for mandatory, 0 for nonmandatory. Choose your approach by how detailed your evaluation needs to be.

The other score to set up is the degree of compliance. In the example given above:

	Relative importance (need)	Vendor compliance
Multiple warehouses	5	
Warehouse hierarchy	3	

Rate vendor compliance scores on the following criteria:

5   Fully compliant with the current system, no customization required

3   Compliant with the current system, customization included by vendor

1   Not compliant without third-party customization

0   The requirement cannot be met

Here, it is better to use a range of scores rather than the binary approach. Why? Vendors can often supply the needed requirement with minor customization. In this case, they would score 3, and you need to be able to distinguish between minor, vendor-provided customization, and significant third-party add-ons. When you issue the RFP, give vendors the above list as well as the degree-of-need scores, so that completed bid

Table 4-3 Sample Categories of Requirements

Category	Subcategory	Sub-subcategory/issue
Operations		
Data analysis	Drill down, graphical, history	
	Work management	Asset hierarchies
		Blanket WOs, approvals, resources, scheduling, safety, crew certification, contractors, condition reporting
		Can Labor hours charged to a Work Order be broken down into regular and overtime hours?
	Preventive maintenance	*Can the system trigger an alarm when equipment's inspection measurements are trending outside a user-defined criterion?*
	Inventory	Reordering, vendor catalogs, multiple warehouses, repairable spares, multiple part numbers, ABC support, service-level costs
		Can the system support multiple warehouses?
		Is a warehouse hierarchy supported?
	Procurement	
	Resources	

Financial	Electronic data collection	*How would bar-coding support issues and receipts?*
	Reporting	*How does the system use and generate a cycle-count report?*
	Accounting methods	
Technical	Concurrent users, numbers of licenses, "power" vs. casual users, architecture, scalability, performance, security/audit logs, databases supported, integration with other systems, data import/export, workflow solution, application architecture, database management, client configurations, development tools, interfaces, capacity performance	*Does the system use constraints (also cascading)?*
Human	Documentation	
	Training	
	User interface	*Can a user have multiple screen access? If so, how?*
	Services	

sheets have two scores for each requirement. When evaluating the bids, multiply both scores to get a "raw" score for the requirement.

What about nonfunctional issues? Vendor track record, support, financial stability, non-fixed-price arrangements, implementation partners? These are as important as functional requirements, sometimes even more so. You need a scoring scheme to compare vendor offerings in these areas. Check with your organization's procurement department. It should have guidelines to follow, and standard scoring rules.

4.3.2.5 Prepare/Issue the Request for Proposal

The RFP is not only a system specification of the CMMS' functional requirements. Depending on your organization's procurement practices, there are also standard terms and conditions, forms of tender, bid bonds, guarantees, and so on, which you will have to assemble, check, and issue. For public organizations, the RFP issue and management process must conform to procurement rules in that jurisdiction. For example, if one vendor raises a query during the bidding period, you may need to formally issue clarifications to them all. Bid opening can be public, with formal processes to manage appeals.

Once you've issued the RFP, the selection teams should develop demonstration scripts for vendors selected for detailed assessment. Demonstration scripts provide a common basis on which to judge how the candidate systems and vendors operate and perform. Without them, vendors, without any constraining requirements, will naturally showcase the best features of their system and you inevitably wind up with an "apples and oranges" comparison. They should focus on important functional issues that are mandatory and reference back to business process maps, if available. Typically, the scripts should be comprehensive enough to cover two to three days of detailed product demonstrations, a reasonable time for modern CMMS applications.

To summarize, task deliverables are:

- Request for Proposal, reviewed and approved by all involved teams
- Clarifications issued during the proposal preparation period
- Communications from vendors
- Detailed demonstration scripts used during assessment

4.3.2.6 Initial (First Cut) Assessment

The initial assessment is reasonably mechanical. As we mentioned, the scores for each functional requirement are multiplied and the result used as that requirement's raw score from each vendor. Calculate the ideal scores (degree-of-need score times the fully compliant score) to calibrate all bids. If the bid results are significantly lower than the ideal, it doesn't necessarily mean a poor response. Perhaps the requirements list was extremely detailed in areas outside the CMMS market. For this reason, it's a good idea to use an expert consultant to build the system specification. He or she should be extremely familiar with each vendor's product and know whether certain requirements can be easily met.

Tabulate the nonfunctional responses (commercial, financial, etc.) and, if a scoring scheme has been set up, apply initial scores. Often, organizations visually inspect the results and this can lead to interpretation problems. For example, which of the following responses to the track record question is "better"?

- We have 10 installations in your industry sector, three of which match your user count.
- We have six completed installations in your industry sector, each of which matches your user count.

Although trivial, this example illustrates the potential for making decisions based on qualitative assessments.

At this time, distribute the initial results to the selection teams and seek opinions. This isn't always easy. Ideally, the rules for joint decision making should be defined up front as part of the project charter. Does a majority decision carry? Is a majority defined as 50% plus 1 or should it be a significant majority? This kind of critical question should be addressed early on, before the decision must be made.

The output of this task includes:

- A documented initial assessment, reviewed and signed off on by each team lead
- A short list of vendors (we suggest a maximum of four) who will be evaluated in detail, with supporting documentation for their inclusion

4.3.2.7 Detailed Assessment

We suggest a two-stage approach. First, a presentation is made by each of the short-listed vendors, concentrating on product overview, corporate background, financial stability, and ability to deliver high-quality services. Follow this with a detailed scripted demonstration/presentation by the two best vendors, emphasizing both product software and services. Of course, you don't have to limit the demonstrations to two vendors. It isn't unusual for three vendors to be involved. The process is time-consuming, though, and expensive. Weigh the benefits of having more than two vendors involved at this stage against the cost of the extra effort. As we continue describing the detailed demonstration steps, assume, for clarity, that only two vendors are involved.

Invite each short-listed vendor (again, we suggest no more than four) to present its credentials in a three-hour presentation. This is to ensure that the vendor's philosophy and way of doing business is consistent with yours, in key areas such as services, support, and company background. Firmly steer the vendor away from detailed software demonstrations at this stage, to concentrate instead on his or her approach to implementation, experience in the industry sector (manufacturing, resource development, utilities, etc.), training methods, and so on. A typical agenda would include:

- General introduction (15 minutes)
- Company overview (45 minutes)
- Questions from selection team (60 minutes)
- Software demonstration (30 minutes)
- Wind-up and remaining questions (30 minutes)

The duration for software demonstrations can vary enormously. If you have developed complex scripted demonstration requirements, these could take several days to execute rather than hours. To help decide on the two finalists, use a scoring scheme for each of the topics above. In particular, team questions and a response-rating system should be decided on in advance. This is complicated because the answers will be delivered interactively. Also, ensure that each selection team has the same set of expectations from the brief software demonstration. Clearly, this isn't an exact process and will require a lot of discussion to work out.

Based on the presentation and evaluation criteria, select the two best vendors and prepare written justifications. Also, notify the losing bidders, clearly spelling out why they were eliminated. In fact, everyone involved in the process should be advised about who made it to the final selection, and why.

From here on, you begin detailed assessment in earnest. Specific steps include:

- Invite each of the two vendors to a site visit. You want them to better understand your operating needs, collect data to use in the final detailed demonstration, and reflect your processes in the final software review.
- Undertake initial reference checks, simultaneously with the site visit, if you wish. This can include conference calls and/or visits to each reference. You want to ensure that the vendor's information is consistent

with the reference user's experience. References must be chosen carefully; their operating environments must be relevant to yours. You should advise the reference in advance about the nature and length of your call, so that he or she can adequately prepare. Naturally, the vendor isn't included in the reference call or visit.

- Invite each finalist for a detailed presentation and software demonstration, following scripts prepared and supplied in advance. There are two major objectives: to ensure that the software is truly suitable and that the vendor can provide high-quality and effective implementation. Vendors, naturally, will demonstrate software attributes that show their system in the best light. So that they address your needs, predetermine that the demonstrations must reflect how the system will be used in your application. Similarly, you want to know how the system implementation would be delivered, by the vendor or a business partner.

Again, this kind of interactive process is best served by preparing in advance. Selection-team members need a common understanding of what they are looking for and pass/fail criteria for the scripted demonstrations.

A critical part of the evaluation is to analyze the implementation and postimplementation services required, together with the vendor's (or implementation partner's) ability to supply them. Include items such as customer support, system upgrades, training quality, postimplementation training, user group meetings/conferences/Web sites, location, and support quality. The selected vendor should provide a sample implementation plan as part of the final presentation/demonstration, followed by a detailed plan, to be approved by the selection teams before a final contract is awarded.

At this point, the winning vendor will most likely be apparent. Carefully document your justification for this and present it to senior management for ratification. Notify the successful vendor, as well as the second-place candidate, who should also be advised that he or she may be invited to continue the evaluation process if a final agreement can't be reached with the preferred vendor.

What if there isn't a clear winner? You could do a detailed functionality test of the two finalists to filter out the best solution. You develop a test instruction set, based on each of the criteria. Because of the functionality depth in most leading CMMS applications, this is a major task, taking several weeks of detailed analysis. To keep it within reasonable bounds, we recommend that only mandatory functions be included. Consider this optional analysis only if, after the detailed demonstrations, reference checks, site visits, and selection-team discussions, you're still deadlocked over the final choice.

4.3.2.8 Contract Award

Before proceeding to contract, hold final discussions with the successful vendor to clarify all aspects of the proposed scope, pricing, resources, and schedule. While it is unlikely that anything major will be uncovered at this stage, remember that up to now the primary focus has been on functionality. This is your opportunity to deal with other important aspects of the vendor's proposal, which also demand your full attention. Once completed, the next step is usually to issue a purchase order. Then the selection team prepares a detailed selection and justification record.

4.4 SYSTEM IMPLEMENTATION

To effectively implement packaged computer systems, three elements must work together:

- People
  - Willingness to change
  - Role changes (e.g., planners, schedulers)
  - Organization change → reporting line change
  - Training effectiveness
- Processes

- How business is done now
- How business should be done
- Technology
  - Hardware and operating systems
  - Application software
  - Interfaces
  - Data

A well-designed implementation project addresses each element, so that the system will be effective and accepted by users. You can apply the outline steps that follow to most packaged business systems. However, the details apply specifically to CMMS implementations.

4.4.1 Readiness Assessment

Before the implementation teams arrive on site with software and hardware, there are some preparatory steps to take. The first should be to conduct what we term a Readiness Assessment, covering:

- Organization and culture issues. This review asks questions such as: Is this company ready for system and process change? Is there a consistent sense of excitement or is there tangible resistance? Is senior management supportive of the initiative and prepared to act as change agents throughout the implementation?
- Business processes. Are they documented, practiced, and understood? Is process change necessary?
- Technology. Is there a need for remedial work to be done before the system is deployed (network, communications, staffing, etc.)?
- Business case. Is the conclusion understood and are appropriate key performance indicators agreed on? How can we be sure the CMMS is delivering the expected benefits?

- Project team. Has it been formed? Do the members understand their roles and responsibilities? If they are drawn from operational staff, do they have the commitment to deal with project, not operational, issues?

Clearly, some of these topics will have been (or should have been) addressed before the RFP was issued. However, it is good practice for the implementation project manager to review them again.

4.4.2 Implementation Project Organization

Several different user groups are needed to successfully implement an enterprise-level CMMS. If you think about where the CMMS sits, functionally, in your organization, this shouldn't be a surprise. Maintenance certainly is front and center, but other skills and staff—warehouse and inventory, procurement and purchasing, accounting, engineering, and project IT support—are also needed to configure the system. And that's just for a "routine" CMMS!

Figure 4-4 shows the relationship among the project teams.

4.4.3 Implementation Plan

A CMMS implementation generally proceeds with the high-level timeline shown in Figure 4-5. In the following sections we deal briefly with each of the project stages shown in the figure.

4.4.4 Project Initiation and Management

This is an ongoing task lasting for the project's duration. The key activities and deliverables are listed in Table 4-4.

4.4.5 Design and Validation

You may find that developed and documented business processes from previous work provide excellent assistance in selecting a new system. However, if the implementation team

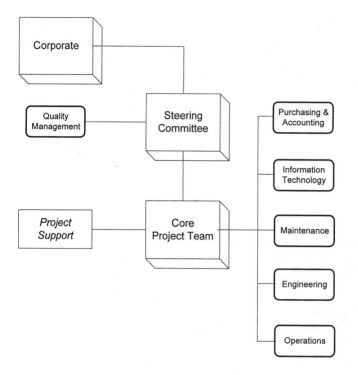

Figure 4-4 Typical project organization.

isn't familiar with them, conduct a review to ensure that they
can be configured into the system. Where there are fit problems, you will have to make some process changes (hopefully
minor—underlining the advantage of having reasonable process maps for the selection teams). Of course, these will be
developed in conjunction with user groups.

The Conference Room Pilot (CRP), also referred to as a
proof of concept configuration, is where all business processes
and user procedures defined earlier are tested and validated.
It is here that you will implement most configuration changes.
The CRP environment is ideal because the data volume is low,
users are knowledgeable, and the impact of configuration
changes is minimal.

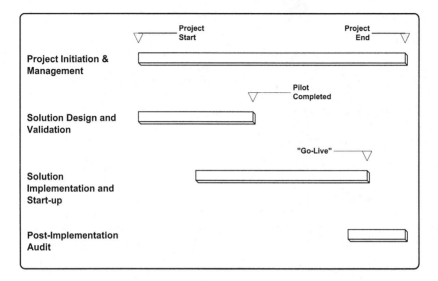

Figure 4-5 Typical CMMS project plan.

Table 4-4   Key Activities and Deliverables in Project Initiation

Activity	Deliverable
Prepare services contract	
Confirm objectives, expectations, Critical Success Factors (CSFs)	Services contract document
Finalize project budget	Project cost report
Prepare project schedule and develop for initial work	Project schedule, expanded to detail level for system configuration and validation
Define and document project procedures (reporting, change control, etc.)	Project procedures document
Detailed activity planning	Updated schedules (rolling wave)
Budget and change control	Change reports and budget/actual reports
Project kick-off meeting	Working project plan

Table 4-5   Key Activities and Deliverables in Design and Validation

Activity	Deliverable
Define and document business processes and user procedures	Business processes document
Configure system with sample data	Conference Room Pilot (CRP) system ready for validation
Provide infrastructure support	Database sizing sheets, network recommendations
Train implementation-team members	Training services and materials
Conduct the CRP	Modified system configuration and associated documentation
Complete the CRP and sign-off	CRP completion document signed off on by the core team members

CRP sign-off, which documents that the system adequately meets the defined functional requirements, typically follows. Also, document shortcomings that can be addressed in subsequent phases to ensure that client-raised issues are tracked and managed throughout the implementation.

The key activities and deliverables are listed in Table 4-5.

4.4.6   Implementation and Startup

It is during this stage that you assemble the production system, having previously validated the base configuration (Table 4-6). Undertake support activities, not directly shown in the table, including infrastructure changes (usually network-, hardware-, and database-related).

The importance of thorough system testing cannot be overstated. It is often inadequate, causing frustration among the user community after go-live. One reason for this is that implementation-team members do not have the time, inclination, or training to develop and conduct detailed test cases. If the staff is available, there is a good argument for bringing in

Table 4-6 Key Activities and Deliverables in Implementation

Activity	Deliverable
Detailed activity/task planning	Updated project schedule
Defined data-conversion requirements	Data-mapping documents
Define interface requirements	Technical design documents, test plans
Develop user training materials	Training package
Develop and implement system testing	System test plan, test results
Deliver user training	Trained users
Final conversion of production data	Converted data on target application
Verify interfaces	Interface sign-offs
Final readiness checks/define resources, etc., disaster planning	Go-live checklists, resource list, contingency plan
Go live	Production system

fresh minds to focus on testing. While this can be expensive, in the long run it's often the cheapest alternative, particularly where there are several integration paths between the CMMS and other systems.

4.4.7 Postimplementation Audit

After initial production operation (go-live), put in place a rigorous monitoring process to ensure that the system is technically stable (performance, availability), being used correctly, and, after a period of operation, producing business benefits (Table 4-7). Although this is the reason that the CMMS was procured in the first place, it is often given scant attention. However, if you set up business measures at the outset of the project, they can easily be measured after an appropriate time.

4.5 CONCLUSION

By now, it must be clear that a CMMS is an indispensable tool for today's Asset Manager. In fact, organizations are looking

Table 4-7   Key Activities and Deliverables in Postimplementation

Activity	Deliverable
Monitor system	Performance reports, database tuning changes, etc.
Obtain user feedback	
Measure achievements against Critical Success Factors (CSFs) and Key Performance Indicators (KPIs)	Analysis of results, where available (may be time-dependent)
Implement required changes, where possible	Configuration changes

at asset management as a core competency, and in many cases the Asset Manager is on a par with the Chief Financial Officer.

The asset-management systems solutions now available cover all organizations, from small single-plant operations to multiplant, multinational companies. These modern systems use the latest technological advances, such as the Internet. Functionally, they are very rich and provide features for the strategic aspects of asset management (long-range budgeting, capital project management) as well as the tactical and operational aspects of maintenance. The latest CMMS advances include built-in "intelligence," to customize maintenance for a particular piece of plant (predictive maintenance). The potential cost savings are huge, without affecting availability. In fact, it too is usually improved.

During the selection process, it's most important to know *what* you are asking for and *why*. In other words, you need to understand the requirements and have a sound business case for buying a CMMS. It isn't unusual for organizations to spend huge sums on asset-management systems and not see any improvement in equipment availability or maintenance costs!

The implementation process is where the "rubber meets the road." Even the best CMMS can be crippled by poor implementation decisions, training, and support. As with any large system, operating a CMMS is a process-driven exercise. It should come as no surprise that if users don't follow the process, the results will be unsatisfactory.

Business processes will always change as a result of industry conditions, personnel issues, and other factors. Recognize that the CMMS is, at heart, just a computer system. Like our automobiles, it needs to be regularly tuned up, with process changes, to operate at maximum efficiency.

BIBLIOGRAPHY

Brodie, Michael, Stonebraker, Michael. Migrating Legacy System: Gateways, Interfaces, and the Incremental Approach. San Francisco: Morgan Kaufman Publishers, 1995.

Linthicum, David S. Enterprise Application Integration. Reading, MA: Addison-Wesley, 1999.

Mylott, Thomas R. III. Computer Outsourcing: Managing the Transfer of Information Systems. Englewood Cliffs, NJ: Prentice Hall Direct, 1995.

Wireman, Terry. Computerized Maintenance Management Systems. New York: Industrial Pr, 1994.

5

Materials Management Optimization

In this chapter, we describe the fundamentals of Maintenance, Repair, and Operations (MRO) materials, focusing on ways to substantially improve your inventory decisions through effective analysis, evaluation, and optimization.

We examine the procurement process through a five-step cycle that includes a discussion of strategic sourcing.

Finally, we look at procurement Internet technology to support the fundamental models developed. Our goal is to help you drive forward MRO management efficiency by quantum steps.

5.1   INTRODUCTION

In most organizations, a lot of money is spent on MRO materials and the expenditures are poorly controlled. Usually, there isn't enough balance-sheet accounting. The result is that inventory isn't optimized or effective procurement strategies for replenishment aren't developed.

The fundamental rationale for storing on site tens of thousands of line items or stock-keeping units (SKUs), costing millions of dollars, is to reduce the Mean Time to Repair for critical equipment. The farther your operation is from parts and supplies distribution centers, the more safety stock is required, and the more critical inventory optimization becomes. The position of the organization on the innocence-to-excellence scale in Figure 5-1 is a good indication of how much the inven-

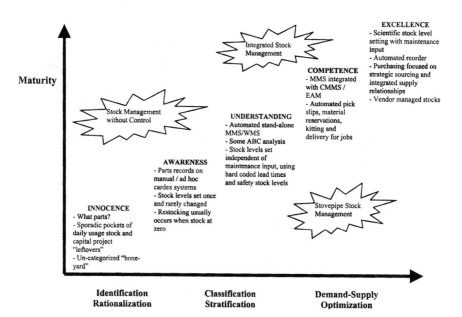

Figure 5-1   Innocence-to-excellence spectrum.

tory is rationalized, inventory control is optimized, and stores purchases are strategically sourced. The lower the position on the scale, the more opportunity to reduce MRO costs and improve service to users—the main job of maintenance, after all.

Using optimization methodologies will help the move from innocence to excellence.

5.2 DEFINING MRO

MRO can be defined as all goods that move through the special inventory account, whether it actually resides in the designated stores or warehouses. This normally doesn't include direct-charge items (goods ordered and charged directly to an operating or maintenance account), materials for manufacturing or processing the product, and items ordered on a capital account. By defining MRO this broadly, you can see the huge impact that the MRO strategy has on the organization. Every procurement decision, from bulldozer parts to tissue paper, must be reviewed and optimized. Table 5-1 shows the characteristics that distinguish MRO from direct materials.

What is most important to understand about MRO is that demand is a random variable. Except for parts used for preventive maintenance (notably for time-based replacement and overhaul), most of the items in the MRO stores are used randomly. In contrast, the consumption rate of tires per automobile manufactured or catalyst used per barrel of oil is highly predictable. For MRO goods, this means that it's often more costly than effective to reorder points and quantity (turnover rate or economic order quantity—EOQ) using traditional materials analysis.

Consider a critical part—a part purchased for a critical piece of production equipment, with a long delivery or lead time. If you use a typical performance criterion for direct materials—high turnover rate to the stocking of this part, say, five turns per year—you might conclude that it's best to remove the part from inventory. But this could expose your operation to potentially costly downtime.

Table 5.1   MRO Defined

	MRO	Direct materials
Usage	Unpredictable long term Dependent on asset life cycle	Predictable over planning horizon Dependent on MRP reliability
Purchasing	Small quantities Dependent on communicated demand (WO Planning, Requisition, direct purchase by end-user) Purchased often	EOQ based quantities (dependent on forecasted demand– MRP) Purchased less frequently to obtain volume discounts
Inventory	Carrying costs usually low (satellite shops, storage yards, end-user defined storage supplements primary storage) Process: receive, store, issue Parts charged to end-use cost center Logical warehouse storage system Turnover rates monitored on small percentage of goods by commodity	Carrying costs are usually high (large volumes require space) Process: receive, issue, WIP, full-goods storage Parts charged to product Bin optimization storage methods All turnover rates monitored
Work management	Asset-centric Resource-driven planning (shutdowns are often scheduled) Unscheduled downtime is not recoverable	Process-centric Capacity/demand driven planning Unscheduled downtime is recoverable if capacity is less than 100%

Applying EOQs could also lead to excessive inventories. Imagine a situation in which for several months you experience repetitive failure on new equipment, subject to high startup stresses. This creates a temporary situation of high use of an expensive part. EOQ calculations inflate the reorder amount. At first the effect may not seem important, but if the part were worth several thousand dollars, or this technique were used extensively, MRO values would be driven excessively high. Consequently, we examine three optimization methods that:

- Optimize inventory by identifying the parts that have been or should be stocked

- Develop inventory control processes that support and sustain optimization decisions
- Strategically source stocked, nonstocked, and one-off purchase items

5.3 INVENTORY OPTIMIZATION

The objective is to optimize your stocking decision by balancing two conflicting cost drivers: stocking materials to minimize stock-out costs versus reducing ownership costs. Using this method, you:

- Identify all MRO sources
- Identify goods that need to be stocked
- Develop new and efficient ways of dealing with goods that should not be stocked

The three steps of inventory optimization are: analysis, evaluation, and optimization.

5.3.1 Analysis: Identification and Rationalization

The first step in optimizing inventories is to analyze current inventory sources and MRO practices. This will help you develop fundamental processes to develop a strategy to reduce costs and optimize stocking decisions.

"Inventory" also includes items that are not held in a warehouse. Often, materials are purchased directly and stored on the shop floor or in designated end-user laydowns (local convenient storage areas in the plant). These inventories are kept for numerous reasons, most often distrust with inventory management, but it's an inefficient practice. It can mean poor stock visibility, inappropriate charges, no assured adherence to specifications or loss protection, and excessive on-hand quantities. All of this is costly to the organization.

You want to consolidate this inventory with all other types into a centralized inventory information source and manage it accordingly. This will help to reduce the amount spent on MRO yet ensure that material is available.

The basic tasks performed at this point are:

Task 1—Identify the entire inventory within the organization, including items that are "off the books."

- List each item's supply and usage dates, by referring to the inventory information sources.
- Identify all inventory held outside the warehouse, including satellite shops, scrap yards, laydown areas, and lockers.
- Construct matrices that segment the inventory according to such criteria as value, transaction frequency, criticality, and likelihood to be stolen, using tools that perform ABC analysis (see Figure 5-2).
- Ensure that all inventory items are uniquely identified and adequately described, and that all components are currently in use.
- Assess common functions and uses across operations, and identify and eliminate duplicate part numbers and sources.

Task 2—Develop a strategy for rationalizing inventories.

- Identify volume and location of material held as inventory.
- Establish a suitable stock-control system, based on inventory size and distribution, transaction volumes, and the integration you need for procurement and work-management systems.

5.3.2 Evaluation: Classification and Stratification

Make stocking decisions first on a commodity level and then, where warranted, at the individual-item level. This ensures not only that the right amount of inventory is available but also that the right types of inventory are stocked and controlled.

To make the appropriate stocking decision, partition the inventory into segments and apply stocking and sourcing strategies to each unit (see Figure 5-2). You will be able to estimate cost savings, using performance measures or industry Key Performance Indicators (KPIs), as described in Chapter 3. Also:

Usage Matrix

Stock Usage Information

		M	R	Overhaul	Operation
Inventory Classification	**A**	- contracts - large quantities highly managed - short PH - low impact	- small purchase / contracts - small quantities highly managed - short – medium PH - med – critical impact	- one-off purchases - very small quantities, highy managed - very long / specialized PH - highly critical impact	- contracts - med – large quantities, medium managed - not in PH - low impact to maintenance – may be critical to operation
	B	- contracts / VMS - large quantities, highly managed - short planning horizon - low impact to downtime	- VMS - small quantities, medium managed - short – medium PH - low impact	- one-off / VMS - rarely purchased, highly managed - specialized PH - medium to critical	- VMS / contracts - large quantities, not managed - not in PH - no impact
	C	- VMS - large quantities, not managed - no PH - little impact	- Not often found	- Not often found	- one off purchases - small quantities, not managed - low PH - low impact

- Purchasing strategies (VMS – Vendor -managed solution
- Inventory Strategies
- Planning horizon (PH)
- Impact to Downtime

- MRO: Maintenance Repair and Operations - Inventory
- Traditionally, the "O" in MRO meant Overhaul
- The above Matrix refers to both options

Figure 5-2 Inventory matrix.

- Establish transactional stock-control and auditing processes, to get a clear stock position and set a control framework for each item.
- Determine stock-level requirements (using statistical stock models such as EOQ, if demand is predictable, or input from the end user if demand is highly variable or seasonal, or varies other than with time) to set optimal levels for individual items.
- Determine how service levels will be calculated and applied.

From Figure 5-3, you can see that there are several factors that apply to each square on the matrix, representing use (inventory classified by usage volume and unit cost) and type (stock-usage information based on why the item was purchased) for each commodity.

If the item you are evaluating is "A class" (the 20% of the inventory that accounts for 80% of the value and/or spend) and was purchased for operations, you could use the following strategies:

Purchasing strategies	Contracts
Inventory strategies	Medium to large quantities, medium amount of stock management (if stocked)
Planning horizon	None
Impact on downtime	None on maintenance; may be critical to daily operation of organization

5.3.3 Optimization

You have now established an optimal stocking level for each commodity and maximum availability (service levels) at minimum cost.

The basic tasks at this step are as follows.

Task 1—Determine the stocking decision for each commodity. There are several stocking choices, depending on whether the item is:

- Regular inventory
- Unique (highly managed) inventory
- Vendor-managed inventory (consignment, vendor-managed at site)
- Vendor-held inventory (the vendor becomes a remote "warehouse" of the organization)
- Stock-less (the item is cataloged and its source identified but not stocked)
- No stock (purchase as required)

Task 2—Increase Maintenance's ability to find the right part for the job. This has become easier now that you have a standard user catalog. The next step is to link each part with the equipment where it is used, by assigning parts in the Bills of Materials (BOM), simplifying the process of identifying material to complete tasks. Accurate and complete BOMs reduce duplicates and are a straightforward, reliable tool for requisitioning materials. You can also use them to accurately determine stock levels. There are two types of BOM:

1. Where-used BOM lists all the repair parts and installed quantities for a piece of equipment or one of its components
2. Job BOM lists all the parts and consumables required for a particular repair job

Task 3—Develop management processes for difficult-to-manage or unique stocks.

- Develop an inventory-recovery (surplus stock) management system, including a decision matrix to help you retain or dispose of surplus stock. This is essential to ensure that surplus stock is managed as an asset to the business. Include a divestment model in this matrix to appraise the immediate disposal return against the probability of future repurchase. Include a plan to identify and value all nonstocked inventories.

- Formalize the repairable-stock management process. Establish accounting practices and policies to effectively control repairable components and stocked materials (e.g., motors, mechanical seals, pump pullout units, and transformers).

5.4 OPTIMIZING INVENTORY CONTROL

Inventory control comprises processes and procedures to support inventory management cost and service level decisions.

- Analyze the current service-delivery methods and the customer requirements. Review all the sites and facilities, and create a basic business process design. This should result in a clear flowchart of all the linked activities forming the process from demand to supply. Review and define information technology and change-management requirements.
- Review and incorporate applicable MRO and inventory-control Best Practices.
- Review performance measurements to determine inventory ownership and set targets and service-level expectations.
- Develop a strategy and implementation plan for the new service-delivery model.

Every organization has constraints and opportunities that have formed its current service-delivery models. It helps, then, to review some of the operating decisions within each process that can have a dramatic effect on optimizing inventory and service levels.

5.5 RECEIVING

Receiving is pivotal to inventory control. There are several key events:

- Debts are paid by matching item quantities and attributes to the purchase order, then confirming with the accounting systems.
- The item as ordered is confirmed.
- The item is marshaled for end use (delivery, storage, inspection).
- All variances or inconsistencies are recorded and monitored and used to track vendor performance.

A major consideration is how much to centralize receiving. You must determine whether trained personnel should handle all receipts at a designated location or whether end users should be in charge of receipts of noninventoried purchases. While either decision has merits, you must review internal factors:

- What is the training/motivation level of the staff? Any employee who receives goods for the organization must implicitly agree to be rigorous, prompt, and accurate.
- Who will actually perform the task? Should the engineer at site receive the goods or the departmental secretary? Who should be fiscally responsible for the task?
- Does centralized receiving of all goods increase the lead time to the end user? Is this a receiving function or an internal communication or systems limitation?

5.6 INVENTORY MAINTENANCE

Maintaining inventory involves identification, storage, and auditing. Inventory audit is the process of confirming the absolute inventory accuracy. Proving inventory accuracy is the equivalent of an operations quality-control program. You must instill and maintain user confidence that whatever the system says is in the warehouse is actually there. Traditional organizations annually "count everything in a weekend," then, as a result, adjust stock balances and value. This method is inefficient, often inaccurate, and usually done under strict time and resource constraints.

Perpetual Stock Count is a better alternative. You set up numerous counts, typically fewer than 200 items, to be counted at regular intervals. Over a year, all inventory (at least A and B class) is counted at least once. The benefits of Perpetual Stocktaking are:

- More accurate counts, as the number of items is relatively small and variances can be easily researched
- Stock identity validated against the description and changes noted, reducing duplicates, increasing parts recognition efficiency, and helping to standardize the catalog
- Confirmed stock location
- Accuracy levels (dollar value and quantity variances) that can be used as a performance measure

5.7 ISSUING AND CUSTOMER SERVICE

Traditionally, tradespeople have turned to stores to search for the "right" part. This has been due largely to a poorly maintained (completely lacking) catalog or parts book. Figure 5-3 represents a typical customer service model.

One goal of optimal customer service is to reduce the time between demand and delivery, or lead time. An example is to replace or reduce issue counter coverage with direct delivery to the end user, reducing tradespeople traffic in stores and using staff for control and auditing.

In developing a service model, service levels are also im-

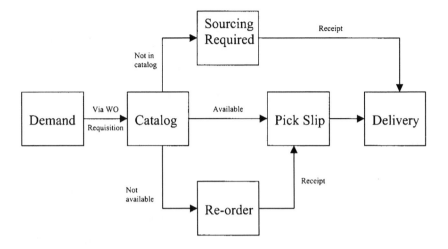

Figure 5-3 Typical demand flow.

portant. Service level (the frequency of stockouts tolerated) is a tool for measuring inventory-control performance. Reduced lead times, inventory accuracy, and the ability to find the right part all contribute to availability. Establishing different service levels for each commodity reduces costs and controls efforts.

5.7.1 Reorder

Automated reorder (for each cataloged item linked to a specific vendor and price) helps reduce lead time and processing. Also, it frees up the purchasing department to negotiate strategic partnerships rather than perform clerical duties.

Reorder is based on reorder points and quantities (maximums/minimums), determined by such factors as historical usage patterns, asset life-cycle analysis, and expected preventive maintenance. The more that reorder is guesswork, the more likely it is that purchasing and carrying costs will increase. By optimizing the reorder process, you reduce costs and increase service levels.

5.7.2 Procurement and Strategic Sourcing

The basic procurement process consists of defining needs, purchasing products and services, receiving them, compensating the supplier, and measuring performance.

Defining needs and purchasing products and services are what's known as sourcing. In this part of procurement, you decide what to buy, how much, from which supplier, and how much to pay. Sourcing is the front end of the traditional, transaction-oriented, purchasing cycle or process. In a relatively uncomplicated enterprise, sourcing is a fairly simple process:

- Define a need for goods or services.
- Locate suppliers.
- Ask for quotes or review the price list.
- Negotiate and establish contracts if necessary.
- Place orders.
- Wait for the goods to arrive.

In strategic sourcing, you clarify and understand the company's need for externally procured goods and services, then match it to supply market capabilities. This moves sourcing from its traditional reactive, transactional function to a much more proactive, strategic role. The basic tasks at this step are listed below.

Task 1—Develop an Initial Demand Management strategy to totally communicate internal planning and scheduling.

- Set service levels appropriate to the customer base (realizing the costs associated with every standard deviation past normal distribution).
- Investigate existing planning horizons and determine sourcing strategies based on the reliability and long-term nature of planned work.
- Reduce overstocked conditions by developing long-term plan reliability.
- Create an automated reorder system.

Task 2—Develop an Initial Supply Management strategy with key suppliers to discuss stockholding requirements, reverse engineering, range rationalization, and demand details.

- Create long-term supplier contracts with fixed pricing, lead times, and service levels (establish the optimal level between accurate lead times and lowest cost). Figure 5-4 shows some of the management issues.
- Automate an order solution for indirect (O-type) MRO materials that reduces sourcing of demands (i.e., E-business).

Task 3—Define the internal and external changes required for the supplier management process, and the roles and responsibilities within the organization to monitor the new supply relationships.

Supplier relationship management approach

Understanding the relationship type and optimizing based on commodity is essential to achieving excellence in procurement

Figure 5-4 Supplier relationships.

- Communicate the supply-chain cost model to other departments and establish the requirements as key elements in tender evaluations and supplier performance assessments.

5.8 PURCHASING OPERATIONS MATERIALS (THE O IN MRO)

We have focused primarily on inventoried materials because of the direct benefits to maintenance—improved material identification, dependable service levels, and reduced costs, to name a few. Now we'll look at how to optimize purchasing operations materials, which is just as important as the inventory-reduction strategies (vendor-managed or held stock) already discussed.

Most purchasing departments identify and monitor much of their procurement performance by measuring cost per purchase order. While this measurement is an effective catalyst for you to understand the impact of purchasing within your organization, it doesn't really address the central issues around inefficient practices. For instance, since most purchasing departments don't grow with the number of orders, they are treated as fixed costs. It's easy to decrease the cost per PO by simply creating single-line purchase orders—increasing the total number of POs. However, this doesn't address such issues as the lack of a consolidated supply base; "maverick," or uncontrolled, departmental spending; or confirmation purchasing—creating a PO from the invoice.

To fully optimize purchasing, you must understand that the core competency of buyers is their ability to procure goods and services as accurately and economically as possible. Buyers, then, must consolidate demand and leverage the volume to reduce acquisition costs and ownership of goods and materials. While this seems to be just common sense, we still see engineers phoning to get quotes and availability information for an electrical breaker from new suppliers. You may have thousands of suppliers on record even if you routinely do business with only 20% of them. What's more, 15–35% of purchase orders for operational spares are for under $1000.

To optimize procurement:

- Identify a strategic source or alternative procurement strategy for each commodity. Set procurement guidelines that are integral to your corporate standardization policy and apply to all departments. Then communicate this plan.
- Instigate better supplier management with a complete supplier-review, negotiation, and consolidation effort based on your sourcing strategies.
- Implement procurement processes that minimize low-value-added activities and provide alternatives to purchase orders (such as the use of procurement cards).
- Invest in technologies and tools that support superior processes, organization, and strategy (order systems that create automatic purchase orders or releases to strategic alliance suppliers).
- Most importantly, develop a procurement culture that encourages problem solving and customer service.

5.9 E-BUSINESS

As the Internet economy accelerates, buying firms target indirect purchasing to cut costs. The opportunity is compelling. For example, if your $2 billion firm reduces the cost of buying processes by 50%, it will save $20 million annually, increasing net profits dollar for dollar. On-line business trade of goods in the United States will reach $1.3 trillion in 2003. Business services likewise have catapulted on-line, becoming "productized" as customers demand better quality control and predictable costs. Recommended pricing levels, service-level agreements, and value-added third-party inspection services ensure quality and efficiency.

Although the phone and fax pioneered instant long-distance communication, their capabilities are severely limited. When companies use services, they interact in many ways: discussing project plans, brainstorming ideas, comparing prices, sending documents, and referencing data. Internet interactions avoid the cost and inconvenience of bringing people into the same room at the same time (1).

Typically, in large organizations procurement- and inventory-management processes have grown out of control. Tens or even hundreds of millions of dollars of indirect spend take place without a unified purchasing system. Using one of the burgeoning automated procurement solutions based on Internet technology, you can:

- Reduce inventory by integrating disparate warehouses, as suppliers provide real-time availability and price
- Reduce spend by expanding the supplier base and eliminating smaller suppliers' entry barriers
- Streamline and simplify procurement while increasing control over purchasing, through automated business rules–based systems that enable distributed desktop requisitioning, internal authorizing, receiving, and payment
- Link buying and supplying organizations into real-time trading communities with instantly accessible services such as descriptive electronic catalogs, requests for quotations, auctions, reverse auctions, shipping and scheduling logistics, order tracking, and payment services
- Reduce paper pushing and increase the level of purchasing value adds such as contract negotiation and supplier management
- Enhance market efficiency by making the buyers' planning and demand forecasts accessible to suppliers

At its most basic, e-business means building electronic links between the organization and its customers and suppliers—connecting companies, departments, customers, and locations. It involves moving processes currently functioning within the business to networks and shared applications. But taken further, e-business is holistic, integrated, and strategy-driven. It goes beyond the front-end, isolated project to demand that we examine, and change, our basic business functions.

E-business is disruptive, forcing companies to rethink their business models. Industry pundits predict increases in e-procurement and e-commerce beyond what most of us can imagine today. Compounding all this is the spate of mergers and acquisitions, which is fueling the need for flexible integrated systems that can consolidate the numerous inventory and procurement systems that autonomous business units previously relied on.

Put simply, an e-business strategy starts with channels such as e-mail, browsers, and shared databases to share information efficiently. Then you link the company to external supplier information about products, pricing, and availability, for instance. Initially, you'll see benefits through quick and efficient information exchange and reduced material lead time.

Once the e-business infrastructure is in place, you can perform more optimization tasks, such as outsourcing non-core business operations or conducting employee training and certification on-line. The common view is that, eventually, e-business will evolve into a network of organizations and markets. Employees gain convenience and choice; the organization benefits from extended, cross-industry value networks. Maintaining your company's relationships, reputation, and unique value proposition with your customers becomes a major priority.

E-business (business-to-business) providers such as ARIBA®, Aspect®, Commerce One®, and MRO.com® are positioning themselves to act as a portal for this network of buyers and suppliers. They will provide the means for buyers to communicate with suppliers even if their output media are different (a business that has EDI will be able to send an order to a business receiving the order through fax). At the organizational level, customers will order the materials and services they need at negotiated prices (or discounted rates).

However, before you consider signing with an e-business technology provider, there are some additional important considerations:

- If the organization hasn't done an MRO Optimization or Business Process Review, it won't achieve any sustainable benefit.
- Does the organization have the infrastructure and Internet capability to support the software?
- Some suppliers are unwilling or unable to accommodate the technical requirements of e-business.
- E-business providers are middlemen (even if you use aggregators). The buying organization must still manage its suppliers. You must ensure that the supplier provides catalogs and that price lists are properly negotiated, as well as monitoring supplier changes communicated through the e-business provider.
- There must be a defined and measurable ROI. Implementing this type of solution is no less resource-intensive than any other enterprise system. And you must also bring smaller suppliers into the equation. All told, it's a very expensive proposition.
- If you have an EAM system, ensure that the functionality you purchased (capture of costs to the asset level or requisitioning through Work Orders) is not lost. Most best-of-breed systems are developing e-business strategies that you will want to explore.

Despite the challenges, these MRO e-business providers offer two significant advantages: they enable your organization to compete in a truly global economy and to reduce costs through strategic sourcing. They also provide a centralized procurement platform for large companies with multiple enterprise systems, offering an alternative to migrating to common enterprise platforms.

5.10   THE HUMAN ELEMENT

Implementing an optimization methodology is essential to achieve the goals of the cost–benefit analysis. However, as with any improvement, unless the policies, practices, and culture that support the organization change to sustain optimization, the benefits will be short-lived.

You must carefully address the "What's in it for me" issue with employees. They must understand their importance in sustaining change and supporting new initiatives. You will need to conduct training and education sessions, but, most importantly, the message must be top-driven. Senior management must communicate its support throughout the organization. The message must be repeated over time to encourage continuous improvement. Conduct performance audits to ensure that optimization is always achieved.

5.11   CONCLUSION

Once you have evaluated your organization's maturity level and settled on long-term goals and expectations, you can apply the appropriate optimization methodologies:

- First, analyze your current inventory and materials sources. Create a sustainable inventory strategy.
- Second, evaluate and adopt inventory control techniques to make optimized stocking decisions and apply performance measurements.
- Third, optimize demand and supply management that

emphasizes strategic sourcing and long-term planning reliability.
- Fourth, review procurement strategies for operations material and implement strategies for better supplier management and spend control.
- Last, realize that change is sustainable only if the employees of the organization are prepared for it. Communication driven from the top down and adequate training are absolutely essential to optimize materials management.

REFERENCE

1. Business services on the net. The Forrester Report. January 1999.

BIBLIOGRAPHY

Leenders, Michiel R., Fearon, Harold E. Purchasing and Supply Management, 11th ed. Chicago: Irwin, 1997.

Nollet, Jean, et al. Operations and Materials Planning and Control. Montreal: G. Morin, 1996.

6

Assessing and Managing Risk

The risks inherent in maintenance management are coming under ever greater scrutiny. In this chapter, we explore key issues in risk management and describe a number of methods to help you assess and manage risk effectively. You will learn the nature of maintenance risk and management processes, including a proven method for identifying critical equipment. Safety factors affecting people and increasing concerns about environmental risk are important areas that will also be dealt with in this chapter.

We also explore potential limitations in using Reliability-Centered Maintenance (RCM) to assess the long-term risks in asset maintenance. It's our view that the current managerial vision of risks should be broadened to include assessments of equipment capability and possible shortcomings in information records and analysis methods. We look at what is needed there, too.

Included are a number of analytical failure/hazard analysis methods that help identify hazards and create risk scenarios. We explore failure modes, effects, and criticality analysis (FMECA) as a way to pinpoint maintenance risks. Similarly, HAZOPS (Hazard and Operability Studies) are reviewed as a comprehensive approach to revealing hazards in an asset's life cycle and determining maintenance, people, and environmental risks. This method has been used in the chemical industry for over 25 years.

Finally, we outline relevant international standards for asset reliability, risk assessment, and risk management that you can use as a reference.

6.1 INTRODUCTION

Maintenance management is about making decisions—about determining the optimal maintenance policy to adopt, whether a piece of equipment needs to be replaced now or can be left in service until the next inspection, how many spares should be held, and so on. While we can try to base these decisions on good data and a rational understanding of the issues and tradeoffs involved, some uncertainty is always involved. It is the nature of our job, therefore, to manage risk.

The term *risk* refers to a situation in which the outcome is uncertain and the consequences generally undesirable. Strictly speaking, then, buying a lottery ticket is a risk, since you don't know in advance whether you will win. But, by this definition, there is no risk associated with jumping off a tall building, since the outcome—instant death—is certain.

Risk is, in effect, the product of probability and consequence. So, two apparently different situations—one with a high probability and low consequence, such as tripping on an uneven floor and hurting your foot, and one with a low probability and high consequence, such as the aircraft you are traveling in crashing and killing everyone on board—can actually have similar risk values.

In maintenance, a joint failure in a compressed-air system causing a minor air leak, for instance, would have relatively small cost and safety consequences, but its risk would be equivalent to that of catastrophic failure of the compressor mainshaft. Although the chance of that happening is very low, it would cause major disruption and be very expensive to repair.

It is only human to focus more on high-consequence events, even when they are unlikely, than on those with little impact. As maintenance engineers, we too can get blindsided in this way. In fact, one of the most significant benefits of Total Productive Maintenance (TPM) is that it addresses conditions such as leaks and minor adjustments that are of minor consequence but happen frequently and deserve our attention.

The rest of this chapter deals with ways to analyze and deal with different risks in a consistent and rational manner.

6.2 MANAGING MAINTENANCE RISK

In any operation, there is always some degree of risk. All activities expose people or organizations to a potential loss of something they value. In maintenance, the impact is typically on equipment failure, people's health and safety, or the environment. Risk involves three issues:

1. The frequency of the loss
2. The consequences and extent of the loss
3. The perception of the loss to the ultimate interested party

A major equipment failure represents an extreme need to manage maintenance risk. The production downtime could delay product delivery to the customer and cost the business a great deal of money. There could be further losses if the equipment failure threatened the safety of employees or adversely affected the environment.

A critical, high-profile failure could also create the impression that the business is out of control and tarnish its reputation in the marketplace. A good example of this is the Perrier water incident a decade ago. A minor maintenance failure led to traces of benzene contaminating the product, which had always promoted an image of purity and health. It was a huge expense to recall and destroy millions of bottles throughout Europe. In addition, the company had to launch a high-profile and expensive public relations campaign to reassure customers that the product was still safe.

Risk can also occur when the wrong maintenance decision is made about an asset. The consequences can be many: lessened plant reliability and availability, reduced product availability, and increased total operating costs.

The best practices for maintaining equipment are based on the premise that more maintenance and prevention result in less downtime, and greater revenues. Figure 6-1 shows that there is an optimal point where the combined preventive and downtime costs are at a minimum. This should determine

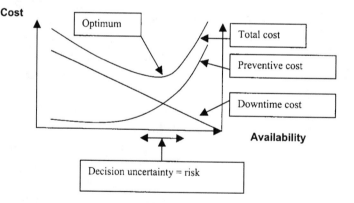

Figure 6-1   Diagram showing the optimal point where the combined preventive and downtime costs are at a minimum.

where you place your maintenance policy and the amount of preventive work you do. While this is a valuable concept, we can never know the exact tradeoff between spending more on preventive maintenance and its impact on downtime. The information that risk decisions are based on will never be totally complete or enable us to foretell the future. Uncertainty is inevitable, so setting the maintenance policy is a question of managing risk.

You can improve the quality of your decisions and be more confident about the optimal point if you have good data. Most risk-management decision processes start by using historical data to predict future events. Generally, that is the only guide we have to the future, based on the premise that history repeats itself. Of course, some factors that affected past events will change and no longer apply. You will still be dealing with uncertainty.

Generally, we don't adequately acknowledge this. If risk is planned for systematically, though, decision making can improve dramatically. To make your process and resulting decisions more credible, you must account for the data you use to make estimates, how you analyze it, and the unknown factors associated with that.

The objective of risk management is to ensure that significant risks are identified and appropriate action taken to minimize them as much as is reasonably possible. To get to this point, you must balance risk-control strategies, their effectiveness and cost, and the needs, issues, and concerns of stakeholders. Communication among stakeholders throughout the process is critical.

Numerous tools and processes can be used to make such decisions. But there is also a tradeoff between the amount of effort put into the analysis and its possible benefit. Getting this wrong can lead to some ridiculous situations, like that of the chemical company that used a full hazard study for a microwave oven in the messroom. The investigators demanded additional interlocks, regular condition monitoring, and radiation monitoring—even though the identical appliance was being used in home kitchens! You can avoid such extreme solutions by seeking help to determine an appropriate risk-management process (1).

There are six general steps in the risk-management process (Figure 6-2):

1. **Initiation**: define the problem and associated risks, form the risk-management team, and identify stakeholders.
2. **Preliminary Analysis**: identify hazards and risk scenarios and collect data.
3. **Risk estimation**: estimate risk frequency and consequences.
4. **Risk evaluation**: estimate benefits, cost, and stakeholders' acceptance of risk.

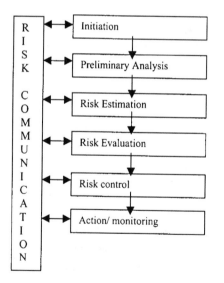

Figure 6-2 The risk-management process broken down into six steps.

5. **Risk control**: identify risk-control options and obtain stakeholders' acceptance of controls and any residual risk.
6. **Action/monitoring**: develop plan to implement risk-management decision and monitor its effectiveness.

To make effective decisions about the risk-management process, the risk team and the stakeholders need to communicate frequently. There must be open dialog so that the team's hypothesis is validated at each step, and to ensure that all stakeholders are involved.

Also, risk assessment is best applied when risks can be readily identified for the critical equipment in a process.

6.3 IDENTIFYING CRITICAL EQUIPMENT

There is generally far more to analyze to improve plant performance than time or resources available. And, as described in the microwave-oven example above, not all equipment deserves the same degree of analysis. To determine where the maintenance improvement effort should go, it is crucial to understand equipment criticality. By this, we mean how critical the piece of equipment is to the business. Generally, the answer is determined by the consequences if it fails.

The elements that make up criticality include safety, health, environment, and financial consequences. Both the financial and nonfinancial impacts can be quantified. Commercial Criticality Analysis (CCA) is valuable to focus and prioritize both day-to-day maintenance and ongoing improvements.

For example, an oil terminal used CCA to rank its 40 main systems for both safety hazards and the cost per hour of downtime. This information was then used by the maintenance foremen to decide which jobs to deal with first. It also helped them determine which units justified flying in spares when a breakdown occurred and which could be served by cheaper, normal delivery. As well, the CCA study helped to refocus the maintenance organization, which had been providing 100% skills coverage, 24 hours a day. A lot of money was saved, without any significant additional risk, by switching to 24-hour coverage only for the critical units. In turn, the CCA findings were used to plan and prioritize a Reliability-Centered Maintenance study, ensuring that it focused on areas of maximum impact.

Assessing criticality is specific to each individual plant. Even within the same industry, what is important to one business may not be to another. Issues such as equipment age, performance, design (a major consideration), technologies using hazardous chemicals, geological issues (typically for mining and oil and gas industries), supplier relationships, product time to market (product cycle), finished-goods inventory policy, and varying national health and safety regulations all influence equipment criticality decisions.

Generally, all businesses need the following equipment-criticality measures:

- Equipment-performance (reliability, availability, and maintainability) measures, as defined in CAN/CSA-Q631-97 (see Section 6.9.1)
- Cost (including direct maintenance and engineering costs and indirect costs of lost production), as defined in BS 6143 Part 2-90 (Section 6.9.6)
- Safety (lost-time incidents, lost-time accidents, etc.)
- Environment (number of environmental incidences, environmental compliance), as defined in ISO 1400— Environmental Management Systems (Section 6.9.5)

6.3.1 Case Study: A Mineral Processing Plant

Let's now examine the top two criticality attributes with a real case study involving a mineral processing plant. Management wanted a maintenance strategy with clear equipment performance and cost measures and a site-wide improvement program.

6.3.1.2 Equipment Performance

Rule One: talk to the people who know the plant. Initially, equipment downtime was based on production statistics, which were clear and accurate to a process line level. But breaking down the picture further required meetings with maintenance and production crews. Note that, as a first analysis, the main concern was equipment causing whole-plant downtime. The downtime statistics are shown in Figure 6-3. The data was presented using Pareto analysis, a simple and ranked presentation style. The conclusion was that the top three equipment types contributed 85% of total plant downtime. Judging from this, conveyors, pumps, and the clarifier were obvious targets for an equipment-performance improvement program.

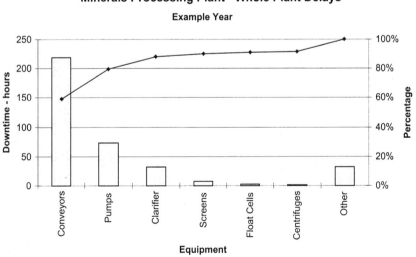

Figure 6-3 Pareto analysis of downtime caused by equipment failures.

The data was then checked with production and maintenance personnel (see Rule One). The downtime to the clarifier was caused by one incident—a piece of mobile plant struck it, causing substantial damage. After investigation, several remedial actions were implemented that should prevent this incident from recurring. The top three targets were then amended to conveyors, pumps, and screens.

The next step was to identify, agree on, and implement conventional Reliability, Availability, and Maintainability (RAM) measures.

6.3.1.3 Costs

The annual operating budget for the plant is shown below:

Operating cost category	Annual spend ($)
Labor—wages	6,300,000
Labor—supervisory	2,100,000
Chemicals	2,500,000
Mechanical parts	4,200,000
Electrical parts	440,000
Energy—electricity	2,300,000
Energy—fuel	210,000
Consultants	150,000
Outside services maintenance	420,000
Contracts maintenance	2,200,000
Other supplies	99,000
Rail freight	5,000,000
Safety services and supplies	280,000
Indirect costs	674,000
Other costs	490,000
Total	27,363,000

Using the existing work order system, costs were then allocated into failure vs. preventive and appraisal categories (a simplified form of BS 6143-1990). Failure costs were compiled from repair and breakdown work orders. High-failure-cost equipment types were identified, scrutinized, and allocated into target subtypes. Failure costs were then extrapolated across the actual maintenance budget, using the work orders as a guide. The results are shown in Figure 6-4.

Using this simple analysis, management identified critical equipment by:

- Poor overall performance
- Failure costs

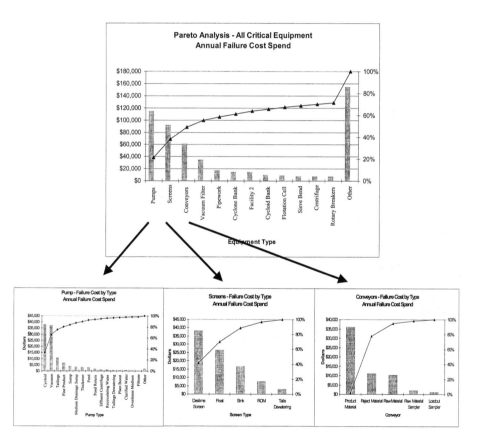

Figure 6-4 Pareto analysis of annual failure-cost spend for all critical equipment.

The data was then used to launch a strategic reliability-improvement program with risk assessment. The organization has since saved over $1 million annually and significantly improved equipment performance, with acceptable risk-management control.

6.4 SAFETY RISK-DUTY OF CARE

Maintenance has a huge impact on safety—the second aspect of risk. As described in this section, the general responsibilities of maintenance engineers can be extensive.

In 1932, Lord Atkin of the British Privy Council ruled that the Stevenson soft drink company was liable for an injury sustained by Scottish widow Mary O'Donahue. This landmark legal precedent became known as the "snail in a bottle" case.

Mrs. O'Donahue and a friend were shopping, and the friend purchased two soft drinks, manufactured by the Stevenson soft drink company. Mrs. O'Donahue partially consumed one of the drinks and later fell violently ill and was hospitalized. It was found that the bottle contained the remains of a decomposed snail. Although Mrs. O'Donahue sought compensation through the Scottish legal system, court after court ruled against her.

Until then, the duty of businesses and citizens to the safety and well-being of others was primarily by contract. Parties who were not contractually obliged were not strictly accountable (i.e., notwithstanding contractual obligations and common law, one party was not accountable for the safety and well-being of another). The courts ruled that Mrs. O'Donahue had no contract with the Stevenson soft drink company. Rather, her friend did because she had purchased the bottle.

Mrs. O'Donahue's lawyer took the matter to the Privy Council, the highest court in the land. After considerable debate, Lord Atkin made his now famous precedent ruling: Regardless of contract, a duty of safety and well-being was owed to Mary O'Donahue.

The ramifications of this ruling 70 years ago include the modern "Duty of Care" legislation on which most national Occupational Health and Safety (OH&S) legislation is based. The ruling ensures that all employers have a duty of care to their employees, employees to one another, and employees to their employer. The standard of care and reasonableness in an organization should follow the national standard. These standards, worldwide, are expected to multiply with time.

6.5 MANAGING LONGER-TERM RISKS

Over the life cycle of an asset, you should seek to maximize reliability and minimize risk. However, certain conditions can increase risk in the long term.

To achieve asset reliability, use failure and reliability analytical methods to develop effective maintenance plans. Failure Modes and Effects Analysis (FMEA), or the closely related Failure Modes, Effects, and Criticality Analysis (FMECA), and Reliability-Centered Maintenance (RCM) use multidisciplinary teams to develop maintenance plans for existing equipment. Although these techniques are invaluable, there are potential long-term risks:

- Slow degradation failures. These are often difficult to predict and model, especially for equipment early in its life cycle. The probability of fatigue-related failures increases with every operating cycle of equipment. It can be difficult, even impossible, to predict these failures using conventional reliability methods. One railway freight provider, aware that low cycle fatigue created increasing failure rates and costs in its wagon fleet, conducted an accelerated reliability test on several wagons. The aim was to model the failures *ahead* of time, so that appropriate condition- and time-based tactics could be developed. Some wagons were fatigue tested to ten times their current lives.

- Incomplete execution of reliability methods. If you don't identify all plausible failure modes, your analysis isn't complete and neither are the maintenance plans. As a result, the equipment will display unpredicted failure modes. You must conduct a periodic review of the plans to evaluate their effectiveness.
- Change in operating environment. Changes in rate of effort (such as operating hours per year), physical environment (such as different geological composition for mining businesses), and operating procedures/techniques can all trigger new failure modes not predicted by the original analysis.
- Change in maintenance environment. Changes in maintenance tactics (such as servicing intervals), maintenance personnel numbers, and/or skill sets, support, and condition-monitoring equipment can lead to unanticipated failures.
- Modifications and capital upgrades. Modifications to existing and/or new equipment can also result in unpredicted failures.

6.6   IDENTIFYING HAZARDS AND USING RISK SCENARIOS

Identifying hazards is the starting point of risk management. You typically do this by analyzing past events, incidents, or lost performance relating to assets, people, and the environment. However, as already mentioned, there are limitations with using past history only. Selecting an experienced team of engineers and process managers is also essential for a thorough understanding of the whole process, the system, and potential hazards or failures.

You develop risk scenarios by identifying hazards and then evaluating the loss, both direct and consequential. A risk scenario is therefore a sequence of events with an associated frequency, probability, and consequences.

You use a variety of approaches and methods to identify risks. Some are based on observation and experiential judgment, others on systematic analysis. Two analytical methods that are typically applied in the asset management field are FMECA and HAZOPS.

6.7 FMEA AND FMECA

The Failure Modes and Effects Analysis (FMEA) identifies potential system failures and their effects. FMEA can be extended to rank failures for their combined severity and probability of occurrence. The failure-ranking process is known as Criticality Analysis (CA). When both steps, FMEA and CA, are performed, the total process is called Failure Mode, Effects, and Criticality Analysis (FMECA).

There are two primary ways to carry out FMEA. One is the hardware approach, which lists the effects on the system. The other is functional, based on the premise that every item in the system is designed to perform a number of functions that can be classified as outputs. For functional FMEA, you list and analyze these outputs to determine their system effects. Variations in design complexity and available data usually dictate the analysis approach to take. When detailed design information is available, you generally study the hardware. The functional approach is most often used in conceptual design stages.

6.7.1 FMEA Objectives

FMEA and FMECA are an integral part of the design process and should be updated regularly to reflect design evolution or changes. FMEA provides inputs to product reviews at various levels of development. Also, FMEA information can minimize risk by defining special test considerations, quality inspection points, preventive maintenance actions, operating constraints, and other pertinent information and activities (such as identifying additional redundancy, alarms, failure-detection mechanisms, design simplification, and derating). FMEA can also be used to:

- Compare various design alternatives and configurations
- Confirm a system's ability to meet its design reliability criteria or requirements
- Provide input data to establish corrective-action priorities and tradeoff studies

6.7.2 FMEA and CA Methodology

FMECA methodology consists of two phases: FMEA and CA. To perform FMEA, you must:

- Define the system and its performance requirements
- Define the assumptions and ground rules you will use in the analysis
- List all individual components or the various functions at the required indenture level for the analysis
- Develop a block diagram or other simple model of the system
- Devise an analysis worksheet to provide failure and effects information for each component, together with other relevant information

CA ranks each potential failure identified in Phase 1 according to its combined severity and probability of occurring. The CA may be performed qualitatively or quantitatively.

You will find the FMEA very versatile and useful for design analysis. If you include the CA, it will be easy to rank failures by their severity and probability of occurrence. Then you will be able to determine the best corrective actions, in order of priority. You can structure the worksheet so that the analysis' detailed information is tailored to fit the situation. The FMEA method, however, has some shortcomings. It can take a great deal of time and effort, making it expensive.

The analysis becomes even more complex if the effects of multiple (two or more simultaneous) failures are taken into account. Both human and external system interactions can easily be overlooked. Often, too much time and effort are spent analyzing failures that have a negligible effect on system performance or safety. You can use a number of computer packages to automate the FMEA analysis. For it to be accurate and effective, though, it should be performed by people intimately familiar with the design and/or process being analyzed (2).

In conclusion, you will find the rigorous FMECA analysis highly effective to detail risk scenarios. Your risk-management team and stakeholders will get a good handle on risk levels and how to control them.

6.8 HAZOPS

HAZOPS, or Hazard and Operability Studies, were developed in the process industry to identify failure, safety, and environmental hazards. They evolved in the chemical industry in the 1970s, particularly under Trevor Kletz, who worked for the United Kingdom–based chemical company ICI. HAZOPS are actually the fourth in a series of six hazard studies covering the life cycle of new plant development, from initial plant concept to commissioning and effective operation. The early studies consider product manufacturing hazards, such as a material's toxicity or flammability, and the pressures and temperatures required. The later studies check that the plant has actually been built according to the requirements of the earlier studies and that operating conditions also comply.

Hazard Study 4 examines in painstaking detail the hazards inherent in a plant's design and any deviations. This is particularly relevant to maintenance. A distressingly high proportion of serious and fatal chemical accidents occur when normal conditions in the plant are temporarily disturbed. Maintenance is important as both an input and an output of the HAZOP—as an input because it can involve changes to normal plant operating conditions and as an output in numerous ways. The study could conclude, for instance, that maintenance be done only under certain conditions, or that some plant items must meet specified standards of performance or reliability. The HAZOP helps determine the importance of various equipment, and the need for either a Commercial Criticality Analysis (CCA) or a FMECA/RCM analysis.

6.8.1 Method

A HAZOP study is carried out by a small team. Typically, it requires an operations expert familiar with the way the plant works (which, in the chemical industry, is probably a chemical engineer), an expert on material hazards (typically, in the chemical industry, a chemist), an expert in the way the plant equipment itself may behave (likely the plant engineer), and a trained facilitator, usually a safety expert. Call in other experts as required to advise on such things as electrical safety and corrosion.

The essential starting point for the HAZOP is an accurate plant diagram (in the process industry, a pipework and instrumentation, or P&I, diagram) and processing instructions spelling out how the product is manufactured in the plant.

Start your study by reviewing the intrinsic manufacturing hazards and standards. This includes briefing the team on the flammability, toxicity, and corrosion characteristics of all the materials, and such matters as high temperatures or pressures, potential runaway reactions, and associated hazards.

You then move on to the painstaking part: walking through the entire manufacturing process, considering each part of the plant down to individual items, even to each section of pipework. For every section considered, you must follow a sequence of prompts. First, you address normal plant operation, to understand each hazard and how it's controlled. Then you have to look at deviations from normal operations. You need a hazard checklist of such things as:

- Corrosion
- Erosion
- Abrasion
- Cracking
- Melting
- Brittleness
- Distortion
- Rupture
- Leakage
- Perforation

You also have to consider what changes could occur in each of the relevant parameters, such as:

- More of
- Less of
- None of
- Reverse of
- Other than

The parameters would include:

- Pressure
- Temperature
- Flow
- pH
- Quantity
- Concentration

As you examine these possibilities, document any hazardous results and the appropriate action to reduce the risk. This could include plant redesign or changes to operating and maintenance procedures. For example, in a plant that relies on automatic safety control systems, more checking may be required to ensure that the instruments are operating correctly. Follow this approach until every part of the plant and the process has been considered and documented.

In summary, you will gain plenty of benefits by using HAZOP. Your plant assessment will be rigorous, comprehensive, and done to high safety standards.

But there are some caveats to keep in mind, too. HAZOP can be time-consuming, typically taking several full days for the team to go through a single process in a complex plant. That kind of thoroughness can be costly, not to mention tedious, if not done efficiently.

6.9 INTERNATIONAL STANDARDS
RELATED TO RISK MANAGEMENT

A number of key standards are described in the following section, describing basic reliability definitions, reliability management, risk management, cost of quality, and software-reliability concepts. These standards are excellent further reading and guidance for Asset Managers.

6.9.1 CAN/CSA-Q631-97—Reliability, Availability, and Maintainability (RAM) Definitions

Scope: This standard lists terms and basic definitions, primarily intended to describe reliability, availability, maintainability, and maintenance support. The terminology is basically that of engineering, but is also adapted to mathematical modeling techniques.

Application: RAM addresses general concepts related to reliability, availability, and maintainability—how items perform to time and under stated conditions, and concerns all the life-cycle phases (concept and definition, design and development, manufacturing, installation, operation and maintenance, and disposal).

Note that in this terminology:

- Reliability, availability, and maintainability are defined as qualitative "abilities" of an item.
- Maintenance support is defined as the qualitative ability of an organization.
- Such general abilities can be quantified by suitable "random variable" conditions, such as "time to failure" and "time to repair."
- You can apply mathematical operations to these random variables using relations and models. The results are called "measures."
- The significance of variables and measures depends on the amount of data collected, the statistical treatment, and the technical assumptions made in each particular circumstance (3).

How to define uptime has been a hot topic of debate within industry over the last decade. Management has spent considerable time grappling with interdepartmental issues, the cost of production downtime vs. maintenance downtime, equipment handover time, logistics, administrative delays, and so on. Often, it has been hard to reach a clear agreement on these issues.

CAN/CSA-Q631-97 contains an excellent series of hierarchical time breakdowns and definitions to help management establish a sustainable measurement process. Figure 6-5, taken from the standard, illustrates this point.

RAM measures are also excellent if you want to establish standard asset-management measures across your organization. This would include individual equipment, process line, systems, and the maintenance organization's ability to support that performance. The clear definition and detail make these measures easily communicated to stakeholders. This reduces the risk of miscommunication and miscomprehension that plagues most managers when establishing an "apples and apples" performance comparison.

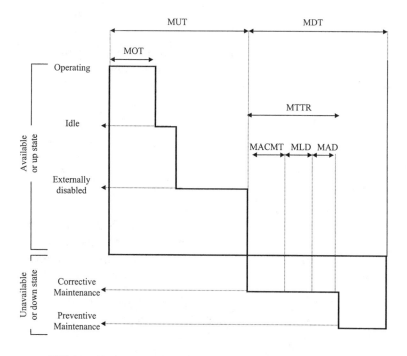

MUT: Mean Up Time   **MACMT**: Mean Active Corrective Maintenance Time
MDT: Mean Down Time   **MLD**: Mean Logistic Delay Time
MTTR: Mean Time To Repair   **MAD**: Mean Administrative Delay Time
MOT: Mean Operating Time

Figure 6-5 Illustration of time-breakdown hierarchy. (From Ref. 3.)

6.9.2 CAN/CSA-Q636-93—Guidelines and Requirements for Reliability Analysis Methods: Quality Management

Scope: This standard guides Asset Managers in selecting and applying reliability-analysis methods. Its purpose is to:

1. Describe some of the most common reliability-analysis methods that represent international standard methods

2. Guide you in selecting analysis methods, depending on the technology and how the system or product is used
3. Establish how results will be documented

Application: No single reliability-analysis method is either comprehensive enough or flexible enough to suit all situations. Consider the following factors to select an appropriate model:

- Analysis objectives and scope
- System complexity
- Consequences of system failure
- Level of detail in design, operation, and maintenance information
- Required or targeted level of system reliability
- Available reliability data
- Specific constraints such as regulatory requirements
- Staff, level of expertise, and resources available

Appendices A to F of the standard contain detailed descriptions of the most common reliability-analysis methods. The following is an overview of how to apply these methods.

Fault Tree Analysis: This method may be suitable when one or more of these conditions apply:

- You need a detailed and thorough system analysis with a relatively high level of resolution
- There are severe safety and economic consequences of a system or component failure
- The reliability requirements are stringent (e.g., system unavailability ≤ 0.001 units?)
- You have considerable staff and resources, including computer facilities

Reliability Block Diagram: Consider this method if one or more of these conditions apply:

- You need either a rudimentary system study or one at a higher hierarchical level (although the method may be used at any level or resolution)
- The system is relatively simple
- You want to keep the analysis simple and straightforward, even if some detail is lacking
- You can obtain reliability data at a block level but data for a more detailed analysis is either not available or not warranted
- The reliability requirements are not very stringent
- You have limited staff and resources

Markov Analysis: This method may be best if one or more of these conditions apply:

- Multiple states or failure modes of the components will be modeled
- The system is too complex to be analyzed by simple techniques such as a reliability block diagram (which may be too difficult to construct or to solve)
- The system has special characteristics, such as:
  A component can't fail if some other specified component has already failed
  You can't repair a component until a certain time
  Components don't undergo routine maintenance if others in the system have already failed

Failure Modes and Effects Analysis (FMEA): The FMEA may be suitable when one or more of these factors apply:

- You need to rank the failure modes' relative importance
- You must detail all possible failure modes along with their effect on system performance

- The system components aren't dependent on each other to any important degree
- The prime concern is single-component failures
- You have considerable resources and staff

Parts Count: Consider this method if one or more of these conditions apply:

- You want to perform a very preliminary or rudimentary conservative analysis
- The system design has little or no redundancy
- You have very limited staff and resources
- The system being analyzed is in a very early design stage
- Detailed information on components such as part ratings, part stresses, duty cycles, and operating conditions are not available

Stress Analysis: You may prefer stress analysis if one or more of the following conditions apply:

- You want to perform a more accurate analysis than the parts-count method
- Considerable staff and resources, including computer facilities, are available
- The system being analyzed is at an advanced design stage
- You have access to detailed information on components such as parts ratings, part stresses, duty cycles, and operating conditions

6.9.3 CAN/CSA-Q850-97—Risk Management: Guidelines for Decision Makers

Scope: The standard (4) helps to effectively manage all types of risks, including injury or damage to health, property, the environment, or something else of value. The standard describes a process for acquiring, analyzing, evaluating, and communicating information for decision making.

Note: The Canadian Standards Association has separate standards to address risk analysis (CSA Standard CAN/CSA-Q634) and environmental risk assessment (CSA Standard Z763)

Application: This standard provides a comprehensive decision process to identify, analyze, evaluate, and control all types of risks, including health and safety. Due to cost constraints, you have to set risk-management priorities, which this standard encourages.

6.9.4   AS/NZS 4360—1999: Risk Management

Scope: The standard helps to establish and implement risk management, including context, identification, analysis, evaluation, treatment, communication, and ongoing monitoring of risks.

Application: Risk management is an integral part of good management practice. It is an iterative process consisting of steps that, in sequence, continually improve decision making. Risk management is as much about identifying opportunities as avoiding or mitigating losses.

This standard may be applied at all stages in the life cycle of an activity, function, project, or asset. You will gain maximum benefits by starting the process at the beginning. It is usual to carry out different studies at various stages of the project.

The standard details how to establish and sustain a risk-management process that is simple yet effective. Here is an overview:

- **Establish the context**: Establish the strategic, organizational, and risk-management context in which the rest of the process will take place. Define criteria to evaluate risk and the structure of the analysis.

- **Identify risks**: Identify what, why, and how problems can arise, as the basis for further analysis.
- **Analyze risks**: Determine the existing controls and their effect on potential risks. Consider the range of possible consequences and how likely they are to occur. By combining consequence and likelihood, you can estimate risk against the pre-established criteria.
- **Evaluate risks**: Compare estimated risk levels against pre-established criteria. That way you can rank them to identify management priorities. If the risks are low, you may not have to take action.
- **Treat risks**: Accept and monitor low-priority risks. For other risks, develop and implement a specific management plan that includes funding.
- **Monitor and review**: Monitor and review how the risk-management system performs, looking for ways to improve it.
- **Communicate and consult**: Communicate and consult with internal and external stakeholders about the overall process, and at individual stages.

The risk-management process is shown in Figure 6-6. Appendix B of the standard details the steps to develop and implement a risk management program:

Step 1	Gain support of senior management
Step 2	Develop the organizational policy
Step 3	Communicate the policy
Step 4	Manage risks at organizational level
Step 5	Manage risks at the program, project and team levels
Step 6	Monitor and review (5)

If you have access to both standards, we encourage you to use them to decide an appropriate process for your environment. Certainly, CAN/CSA-Q850-97 describes very well the

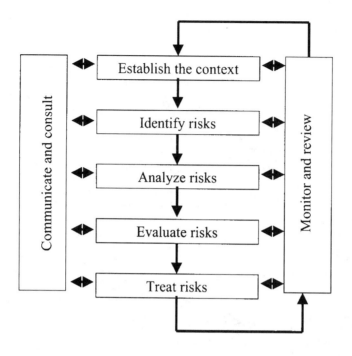

Figure 6-6 The risk-management process.

risk-communication process, including stakeholder analysis, documentation, problem definition, and general communications. AS/NZS 4360—1999, on the other hand, has a well-developed and articulated risk-management process model. AS/NZS 4360—1999 also provides a useful summary for implementing a risk-management program.

6.9.5 ISO 14000: Environmental Management Systems

ISO 14000 is a series of international, voluntary environmental management standards. It is included here because environmental damage is one of the key risks we need to consider in maintenance. Developed by ISO Technical Committee 207, the 14000 series of standards addresses the following aspects of environmental management:

- Environmental Management Systems (EMS)
- Environmental Auditing and Related Investigations (EA&RI)
- Environmental Labels and Declarations (EL)
- Environmental Performance Evaluation (EPE)
- Life-Cycle Assessment (LCA)
- Terms and Definitions (T&D)

The ISO series of standards provide a common framework for organizations worldwide to manage environmental issues. They broadly and effectively improve environmental management, which in turn strengthens international trade and overall environmental performance.

The key elements of an ISO 14001 EMS are:

- **Environmental policy**: the environmental policy and how to pursue it via objectives, targets, and programs
- **Planning**: analyzing the environmental aspects of the organization (including processes, products, and services, as well as the goods and services used by the organization)
- **Implementation and operation**: implementing and organizing processes to control and improve operations that are critical from an environmental perspective (including both products and services)
- **Checking and corrective action**: checking and corrective action including monitoring, measuring, and recording characteristics and activities that can significantly impact the environment

- **Management review**: top management review of the EMS to ensure that it continues to be suitable and effective
- **Continual improvement**: continual improvement is a key component of the environmental management system. It completes the cycle—plan, implement, check, review, and continually improve.

You can obtain final published copies of ISO 14000 standards and related documents from your National Standards Association (ISO Member Body), which is usually a country's primary ISO sales agent. In countries where the national standards association is not an ISO member body, ISO 14000 documents can be obtained directly from the ISO Central Secretariat.

6.9.6   BS 6143—1990: Guide to the Economics of Quality

Scope: BS 6143 is divided into two parts.

Part 1   Process cost model: Using this model, process measurement and ownership are key, and you can apply quality costing to any process or service. The quality cost categories simplify classification by making clear the cost of conformance and nonconformance. The method involved is process modeling, and there are guidelines for various techniques. In addition, the process control model is compatible with total quality management.

Part 2   Prevention, appraisal, and failure model: This is a revised version of traditional product quality costing in manufacturing industries. With recent improvements, this approach has become more effective, although you may want to combine it with the process cost model.

Using this standard will help you determine the cost of preventing defects, appraisals, and internal and external failures, as well as quality-related cost systems for effective business management.

Application: For Asset Managers unfamiliar with this standard, it deals with a manufacturing cost structure that can readily be applied to direct maintenance charges. Costs are defined as follows:

Prevention cost: the cost of any action to investigate, prevent, or reduce the risk of nonconformity or defect.

Appraisal cost: the cost of evaluating quality requirement achievements, such as verification and control performed at any stage of the quality loop.

Internal failure cost: the cost of nonconformities or defects at any stage of the quality loop, including, for example, scrap, rework, retest, reinspection, and redesign.

External failure cost: the cost of nonconformities or defects after delivery to a customer/user. This can include claims against warranty, replacement, and consequential losses, as well as evaluating penalties.

Identifying cost data: quality-related costs should be identified and monitored. It is essential that the way you classify the data be relevant and consistent with other accounting practices within the company; otherwise you won't be able to fairly compare costing periods or related activities.

Quality-related costs are a subset of business expenses, and it's useful to maintain a subsidiary ledger or memorandum account to track them. By using account codes within cost centers, you will better monitor the quality cost of individual activities. Allocating costs is important to prevent failures and it shouldn't be done solely by an accountant. You may need technical advice as well.

Quality costs alone don't provide managers with enough perspective to compare them to other operating costs or identify critical problem areas. To understand how significant a quality cost actually is, compare it with other costs in the organization that are regularly reported (6).

6.9.7 ANSI/AIAA R-013—1992: Software Reliability (American National Standard)

Scope: Software Reliability Engineering (SRE) is an emerging discipline that applies statistical techniques to data collected during system development and operation. The purpose is to specify, predict, estimate, and assess how reliable software-based systems are. This is a recommended practice for defining software reliability engineering.

Application: The techniques and methods in this standard have been successfully applied to software projects by industry practitioners in order to:

- Determine whether a specific software process is likely to produce code that satisfies a given software reliability requirement
- Determine how much software maintenance is needed by predicting the failure rate during operation
- Provide a metric by which to evaluate process improvement
- Assist software safety certification
- Determine whether to release a software system or to stop testing it
- Estimate when the next software system failure will occur
- Identify elements in the system that most need redesign to improve reliability
- Measure how reliably the software system operates, to make changes where necessary

Basic Concepts: There are at least two significant differences between hardware and software reliability. First, software does not fatigue, wear out, or burn out. Second, because software instructions within computer memories are accessible, any line of code can contain a fault that could produce a failure.

The failure rate (failures per unit time) of a software system, as shown in Figure 6-7, is generally decreasing due to fault identification and removal.

Procedure: A 13-step generic procedure for estimating software reliability is listed below. This should be tailored to the project and the current life-cycle phase. Not all steps will be used in every application, but the structure provides a convenient and easily remembered standard approach. The following steps represent a checklist for reliability programs:

1. Identify application.
2. Specify the requirement.
3. Allocate the requirement.
4. Define failure—a project-specific failure definition is usually negotiated by the testers, developers, and users. It is agreed on before the test begins. What

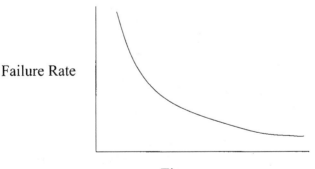

Failure Rate

Time

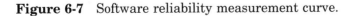

Figure 6-7 Software reliability measurement curve.

is most important is that the definition be consistent over the life of the project.

5. Characterize the operational environment—this includes three aspects: system configuration, system evolution, and system operating profile.

6. In modeling software reliability, keep in mind that systems frequently evolve during testing. New code and components can be added.

7. Select tests—software-reliability engineering often involves operations and collecting failure data. Operations should reflect how the system will actually be used. The standard includes an appendix of information to help determine failure rates.

8. Select models—included are various reliability models. We recommend that you compare several models before making a selection.

9. Collect data—to make an effective reliability program, learn from previous lessons. This doesn't mean you need to keep every bit of information about the program as it evolves. Also, you must clearly define data-collection objectives. When a lot of data is required, it's going to affect the people involved. Cost and schedule can suffer too.

10. Two additional points to keep in mind when collecting data: motivate the data collectors and review the collected data promptly. If you don't follow this advice, quality will suffer.

11. Estimate parameters—there are three techniques in the standard to determine model parameters: method of moments, least squares, and maximum likelihood.

12. Validate the model—to properly validate a model, first address the assumptions about it. You can effectively do this by choosing appropriate failure data items and relating specific failures to particular intervals or changes in the life cycle.

13. Perform analysis—once you have collected the data and estimated the model parameters, you're ready to perform the appropriate analysis. Your objective may be to estimate the software's current reliability, the number of faults remaining in the code, or when testing will be complete.

Be careful about combining a software-reliability value with your system-reliability calculation. The analysis may call for a system-reliability figure while the software reliability is calculated in execution time. In that case, it must be converted to calendar time in order to be combined with hardware reliabilities. By doing this, you will be able to calculate the system value (7).

REFERENCES

1. CAN/CSA-Q850-97. Risk Management: Guideline for Decision Makers.
2. CAN/CSA/Q636-93. Quality Management: Guidelines and Requirements for Reliability Analysis Methods.
3. CAN/CSA-Q631-97. Reliability, Availability and Maintainability (RAM) Definitions.
4. CAN/CSA-Q850-97. Risk Management: Guidelines for Decision Makers.
5. AS/NZS 4360—1999. Risk Management.
6. BS 6143—1990. Guide to the Economics of Quality.
7. ANSI/AIAA R-013—1992. Software Reliability (American National Standard).

7

Reliability by Design: Reliability-Centered Maintenance

Reliability-Centered Maintenance (RCM) is well established as the pre-eminent technique for establishing a scheduled maintenance program. In this chapter, we introduce RCM, describe it in detail, and explore its history. We discuss who should be using RCM and why. RCM is increasingly important as society becomes more litigious and productivity demands increase.

The entire RCM process is described, along with important factors to consider as you work through an RCM analysis. The chapter includes a flow diagram for a suggested process that complies with the new SAE standard for RCM programs, and a simplified decision logic diagram for selecting appropriate and effective maintenance tactics. We explain the deliverables and how to get them from the vast amount of data that is usually produced by the process. The scope of RCM projects is also described, so that you can get a feel for the effort involved. We thought it would be helpful to include an effective RCM implementation that shows team composition, size, time and effort required, and tools that are available to make the task easier.

Of course, RCM has been used in several environments and has evolved in numerous ways. Some are only slight variations of the thorough process, others are less rigorous, and some are downright dangerous. These methods are discussed along with their advantages and disadvantages. We also examine why RCM programs fail, and how to recognize and avoid those problems. As responsible maintenance and engineering professionals, we all want to improve our organization's effectiveness. You will learn how to gradually introduce RCM, successfully, even in the most unreceptive environments.

This chapter is likely to generate some controversy and discussion. That is just what we intend. In law, the concept of justice and the legal realities are sometimes in conflict. Similarly, you will find that striking a balance between what is right and what is achievable is often a difficult challenge.

7.1   INTRODUCTION

As mentioned above, RCM is the pre-eminent method for establishing a scheduled maintenance program. For years it has been demonstrated to be highly effective in numerous industries—civil and military aviation, military ship and naval weapon systems, electric utilities, and the chemical industry. It is mandated in civil aircraft and often, as well, by government agencies procuring military systems. Increasingly, RCM is being selected when reliability is important for safety or en-

vironmental reasons or simply to keep a plant running at maximum capacity.

The recently issued SAE Standard JA1011, "Evaluation
Criteria for Reliability-Centered Maintenance (RCM) Processes," (1) outlines the criteria a process must meet to be
called RCM. This new standard determines, through seven
specific questions, whether a process qualifies as RCM, although it doesn't specify the process itself.

In this chapter, we describe a process designed to satisfy
the SAE criteria. Plus, we've included several variations that
don't necessarily answer all seven questions but are commonly
called RCM. Consult the SAE standard for a comprehensive
understanding of the complete RCM criteria.

In our increasingly litigious society, we are more and
more likely to be sued for "accidents" that may be caused by
events or failures previously accepted as being out of our control. Today, the courts take a harsh stand with those who
haven't done all they could to eliminate risks. There are many
examples in recent decades of disastrous "accidents" that
could have been avoided, such as the carnage at Bhopal, the
Challenger explosion, and the tendency of the original Ford
Pinto gas tanks to explode when rear-ended.

The incident at Bhopal triggered sweeping changes in the
chemical industry. New laws were established, including the
Emergency Planning and Community Right to Know Act
passed by the U.S. Congress in 1986 and the Chemical Manufacturers Association's "Responsible Care" program. Following the Pinto case and others of a similar nature, consumer
goods manufacturers are being held to ever more stringent
safety standards.

Despite its wide acceptance, RCM has been criticized as
being too expensive just to solve the relatively simple problem
of determining what maintenance to do. These criticisms often
come from those in industries where equipment reliability and
environmental compliance and safety are not major concerns.
Sometimes, as well, they result from failing to manage the

RCM project properly, as opposed to flaws within RCM itself. Alternative methods to RCM are covered later in this chapter, although we don't recommend that you use them. A full description of their risks is included.

We all want to get the maximum from our transportation and production systems and plants. They're very expensive to design and build, and downtime is costly. Some downtime is needed, of course, to sustain operations and for logical breaks in production runs or transportation schedules. RCM helps eliminate unnecessary downtime, saving valuable time and money.

RCM generates a scheduled maintenance program that logically anticipates specific failure modes. It can also effectively:

- Detect failures early enough to be corrected quickly and with little disruption
- Eliminate the cause of some failures before they happen
- Eliminate the cause of some failures through design changes
- Identify those failures that can safely be allowed to occur

This chapter provides an overview of the different types of RCM:

- Aircraft vs. military vs. industrial
- Functional vs. hardware
- Classical (thorough) vs. streamlined and "lite" versions

We describe the basic RCM process, step by step. This includes a brief overview of critical equipment and FMECA, which are covered thoroughly in Chapter 6 ("Assessing and Managing Risk").

7.2   WHAT IS RCM?

RCM is a logical, technical process that determines which maintenance tasks will ensure a reliable design system, under specified operating conditions, in a specified operating environment. Each of the various reference documents describing RCM applies its own definition or description. We refer readers to SAE JA1011 (1) for a definitive set of RCM criteria.

RCM takes you from start to finish, with well-defined steps arranged in a sequence. It is also iterative—it can be carried out a few different ways until initial completion. RCM determines how to improve the maintenance plan, based on experience and optimizing techniques. As a technical process, RCM delves into the depths of how things work and what can go wrong with them. Using RCM decision logic, you select maintenance interventions or tasks to reduce the number of failures, detect and forecast when one will be severe enough to warrant action, eliminate it altogether, or accept it and run until failure.

The goal of RCM is to make each system as reliable as it was designed to be. Each component has its own unique combination of failure modes and failure rates. Each combination of components is also unique, and failure in one component can cause others to fail. Each "system" operates in its own environment, consisting of location, altitude, depth, atmosphere, pressure, temperature, humidity, salinity, exposure to process fluids or products, speed, and acceleration. Depending on these conditions, certain failures can dominate. For example, a level switch in a lube oil tank will suffer less from corrosion than one in a saltwater tank. An aircraft operating in a temperate maritime climate is likely to corrode more than one in an arid desert. The environment and operating conditions can have significant influence on what failures will dominate the system.

It is this impact of operating environment on a system's performance and failure modes that makes RCM so valuable.

Technical manuals often recommend a maintenance program for equipment and systems, and sometimes they include the effects of the operating environment. For example, car manuals specify different lubricants and antifreeze densities that vary with ambient operating temperature. However, they don't usually address the wear and tear of such factors as driving style (aggressive vs. timid) or how the vehicle is used (taxi or fleet vs. weekly drives to visit family). In industry, manuals are not often tailored to any particular operating environment. An instrument air compressor installed at a subarctic location may have the same technical manual and dewpoint specifications as one installed in a humid tropical climate. RCM specifically addresses the environment experienced by the fleet, facility, or plant.

7.3   WHY USE RCM?

Because it works.

RCM has been around for about 30 years, since the late 1960s, beginning with studies of airliner failures carried out by Nowlan and Heap (2). United Airlines wanted to reduce the amount of maintenance for what was then the new generation of larger wide-bodied aircraft. Previously, aircraft maintenance was based on experience and, because of obvious safety concerns, it was quite conservative. As aircraft grew larger, with more parts and therefore more things to go wrong, maintenance requirements similarly grew, eating into flying time needed to generate revenue. In the extreme, achieving safety could have become too expensive to make flying economical. But, thanks to RCM and United's willingness to try a new approach, the airline industry has been able to develop almost entirely proactive maintenance. The result? Increased flying hours, with a drastically improved safety record.

In fact, aircraft safety has been consistently improving since RCM was introduced. In addition, RCM has reduced the number of maintenance man-hours needed for new aircraft per flight. Why? RCM identifies functional failures that can be caught through monitoring before they occur. It then reveals which failures require some sort of usage- or time-based intervention, develops failure-finding tests, and indicates whether system redesign is needed. Finally, it flags failures that can be allowed to occur because they cause only minor problems— where aircraft are concerned, this is a very small number indeed. Frequent fliers seldom experience delays for mechanical or maintenance-related problems, and airlines are usually able to meet their flight schedules.

RCM has also been used successfully outside the aircraft industry. Those managing military capital equipment projects, impressed by the airlines' highly reliable equipment performance, often mandate the use of RCM. The author participated in a shipbuilding project in which total maintenance workload for the crew was almost 50% less than for other similarly sized ships. At the same time, the ship's service availability improved from 60% to 70%. The amount of downtime needed for maintenance was greatly reduced. In that project, the cost of performing RCM was high—in the millions of dollars—but the payback, in hundreds of millions, justified it.

The mining industry usually operates in remote locations far from sources of parts, materials, and replacement labor. Consequently, miners, like the Navy and the airline industry, want high reliability, minimum downtime, and maximum productivity from the equipment. RCM has been a huge benefit. It has made fleets of haul trucks and other equipment more available, while reducing maintenance costs for parts and labor and planned maintenance downtime.

In process industries, RCM has been successful in chemical plants, oil refineries, gas plants, remote compressor and pumping stations, mineral refining and smelting, steel, aluminum, pulp and paper mills, tissue-converting operations, food and beverage processing, and breweries. RCM can be applied in any operation where high reliability and availability are important.

7.4   WHO SHOULD USE RCM?

Any plant, fleet, or building in which productivity is crucial. That includes companies that can sell everything they produce, where uptime and high equipment reliability and predictability are very important. It also includes anyone producing to meet tight delivery schedules, such as just-in-time parts delivery to automotive manufacturers, where equipment availability is critical. Availability means that physical assets (equipment, plant, fleet, etc.) are there when needed. The higher the equipment availability, the more productive the assets. Availability (A) is measured by dividing the total time that the assets are available by the total time needed for them to run.

If a failure causes damages, there's a growing trend in our increasingly litigious society for those affected to sue. Practically everyone, then, can benefit from RCM. There is no better way of ensuring that the right maintenance is being done to avoid or mitigate failures. Availability can be expressed as:

$$A = \frac{Uptime}{Total\ time}$$

Uptime is simply total time minus downtime. Downtime for unplanned outages is the total time to repair the failures, or Mean Time to Repair (MTTR) multiplied by the number of failures. Reliability takes into account the number of unplanned downtime incidents suffered. Reliability is, strictly speaking, a probability. A commonly used interpretation is that large values of Mean Time Between Failures (MTBF) indicate highly reliable systems. Generally, plants, fleets, and buildings benefit from greater MTBF because it means fewer disruptions.

RCM is important in achieving maximum reliability—longest MTBF. For most systems, MTBF is long and, typically, repair time (MTTR) is short. Reducing MTTR requires a high level of maintainability. Adding the two gives the total time that a system could be available if it never broke down. Availability is the portion of this total time that the system is actually in working order and available to do its job.

Availability can also be written as:

$$A = \frac{MTBF}{MTBF + MTTR}$$

Mathematically, we can see that maximizing MTBF and minimizing MTTR will increase A.

Generally, an operation is better off with fewer downtime incidents. If downtime is seriously threatening the manufacturing process, delivery schedule, and overall productivity, additional measures clearly are needed. Reliability is paramount. RCM will help you maximize MTBF while keeping MTTR low.

7.5   THE RCM PROCESS

RCM has seven basic elements to meet the criteria of the new SAE standard:

1. Identify the equipment/system to be analyzed.
2. Determine its functions.
3. Determine what constitutes failure of those functions.
4. Identify what causes those functional failures.
5. Identify their impacts or effects.
6. Use RCM logic to select appropriate maintenance tactics.
7. Document the final maintenance program and refine it as operating experience is gained.

Items 2 through 5 constitute the Failure Modes and Effects Analysis (FMEA) portion of RCM, discussed in greater detail in Chapter 6. Some practitioners limit their FMEA analysis only to failures that have occurred, ignoring those that may have been effectively prevented. But FMEA, as used in the RCM context, must consider all possible failures—ones that have occurred, can currently be prevented, and haven't yet happened.

In the first step of RCM, you decide what to analyze (Figure 7-1). A plant usually contains many different processes, systems, and types of equipment. Each does something different, and some are more critical to the operation than others. Some equipment may be essential for environmental or safety reasons but have little or no direct impact on production. For example, if a wastewater effluent treatment system that prevents untreated water being discharged from the plant goes down, it doesn't stop production. The consequences can still be great, though, if environmental regulations are flouted. The plant could be closed and the owners fined or jailed.

Start by establishing criteria to determine what is important to the operation. Then, use them to decide which equipment or systems are most important, demanding the greatest attention. There are many possible criteria, including:

- Personnel safety
- Environmental compliance
- Production capacity
- Production quality
- Production cost (including maintenance costs)
- Public image

When a failure occurs, the effect on each of these criteria can vary from "no impact" or "minor impact" to "increased risk" or "major impact." Each criterion and how it is affected can be weighted. For example, safety usually rates higher than production capacity. Likewise, a major impact is weighted higher than no impact. For each system or piece of equipment

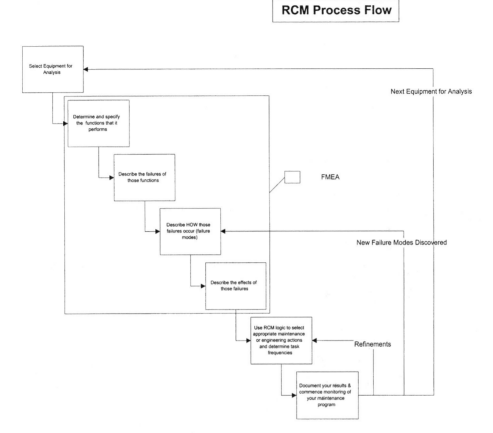

Figure 7-1 RCM process overview.

being considered for RCM analysis, imagine a "worst case" failure, then determine its impact on each criterion. Multiply the weights of each criteria and add them to arrive at a "criticality score." Items with the highest score are most important and should be analyzed first.

There are both active and passive functions in each system. Active functions are usually the obvious ones, for which we name our equipment. For example, a motor control center controls the operation of various motors. Some systems also have less obvious secondary or even protective functions. A chemical process loop and a furnace, for instance, both have a secondary containment function. They may also include protective functions such as thermal insulating or chemical corrosion resistance.

Keep in mind that some systems, such as safety systems, are normally dormant and do not become active until some other event has occurred. Unfortunately, normally passive-state failures are often difficult to spot until it's too late.

Each function also has a set of operating limits, defining the function's "normal" operation and its failures. When the system operates outside these "normal" parameters, it has failed. Our system failures can be categorized in various ways, such as: high, low, on, off, open, closed, breached, drifting, unsteady, and stuck. Remember that the function fails when it falls short of or exceeds its operating environment's specified parameters.

It is often easier to determine functions for the individual parts than for the entire assembly. There are two ways to analyze the situation. One is to look at equipment functions, a fairly high assembly level. You must imagine everything that can go wrong. This works well for pinpointing major failure modes. However, you could overlook some less obvious possible failures, with serious consequences.

An alternative is to look at "part" functions. This is done by dividing the equipment into assemblies and parts, similar to taking the equipment apart. Each part has its own functions and failure modes. By breaking down the equipment into parts, it is easier to identify all the failure modes, without missing any. This is more thorough, but it does require more work.

To save time and effort, some practitioners perform Pareto analysis on the failure modes to filter out the least common ones. In RCM, though, all failures that are reasonably likely should be analyzed. You must be confident that a failure is unlikely to occur before it can be safely ignored.

A failure mode is physical. It shows *how* the system fails to perform its function. We must also identify *why* the failure occurred. The root cause of failures is often a combination of conditions, events, and circumstances.

For example, a hydraulic cylinder may be stuck in one position. The cylinder has failed to stroke or provide linear motion. *How* it fails is the loss of lubricant properties that keep the sliding surfaces apart. There are many possibilities, though, for *why*. The cause could be a problem with the fluid— using the wrong one, leaking, dirt, or surface corrosion due to moisture. Each of these can be addressed by checking, changing, or conditioning the fluid.

Not all failures are equal. They can have varying effects on the rest of the system, plant, and operating environment. The cylinder's failure in the above example could be severe if, by actuating a sluice valve or weir in a treatment plant, excessive effluent flows into a river. Or the effect could be as minor as failing to release a "dead-man" brake on a forklift truck stacking pallets in a warehouse. In one case, it's an environmental disaster, in the other, only a maintenance nuisance. But if an actuating cylinder on the brake fails in the same forklift while it's in operation, there could be serious injury.

By knowing the consequences of each failure we can determine what to do and whether it can be prevented, predicted, avoided altogether through periodic intervention, eliminated through redesign, or requires no action. We can use RCM logic to choose the appropriate response.

RCM helps to classify failures as hidden or obvious, and whether they have safety, environmental, production, or maintenance impacts. These classifications lead the RCM practitioner to default actions if appropriate predictive or preventive measures cannot be found. For example, a fire-sprinkler failure can't be detected or predicted while it is in

Figure 7-2   Normally, a fire sprinkler is "dormant."

normal operation (dormant), but by designing in redundancy
of sprinklers we can mitigate the consequences of failure (Figure 7-2). More severe consequences typically require more extensive mitigating action. The logic is depicted in Figure 7-3.

Most systems involving complex mechanical, electrical,
and hydraulic components will have random failures. You
can't confidently predict them. Still, many are detectable before functional disruption takes place. For example, if a
booster pump fails to refill a reservoir providing operating
head to a municipal water system, the system doesn't have to
break down. If you watch for the problem, it can be detected
before municipal water pressure is lost. That is the essence of
condition monitoring. We look for the failure that has already
happened but hasn't progressed to the point of degrading the
system. Finding failures in this early stage helps protect overall functional performance.

Since most failures are random in nature, RCM logic first
asks if it is possible to detect the problem in time to keep the
system running. If the answer is "yes," condition monitoring
is needed. You must monitor often enough to detect deterioration, with enough time to act before the function is lost. For
example, in the case of the booster pump, check its performance once a day if you know that it takes a day to repair it
and two days for the reservoir to drain. That provides a buffer

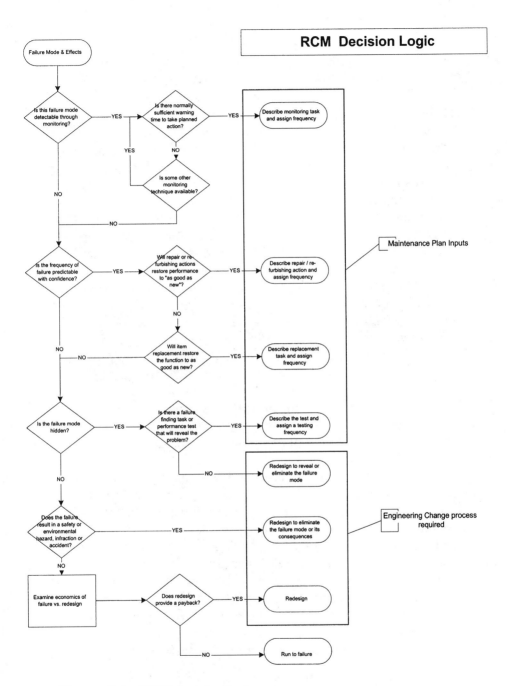

Figure 7-3 RCM decision logic.

of at least 24 hours after detecting and solving the problem before the reservoir system is adversely affected.

If you can't detect the problem in time to prevent failure, RCM logic asks if it's possible to reduce the impact by repairing it. Some failures are quite predictable even if they can't be detected early enough. For example, we can safely predict brake wear, belt wear, tire wear, erosion, etc. These failures may be difficult to detect through condition monitoring in time to avoid functional failure or they may be so predictable that monitoring for the obvious isn't warranted. After all, why shut down monthly to monitor for belt wear if you know it isn't likely to appear for two years? You could monitor every year but, in some cases, it's more logical to simply replace the belts automatically every two years. There is the risk, of course, that a belt will either fail earlier or still be operating well when it's replaced. If you have sufficient failure history you may be able to perform a Weibull analysis to determine whether the failures are random or predictable.

If the above approach isn't practical, you may need to replace all the equipment. Usually, this makes sense only in critical situations because it typically requires an expensive sparing policy. The cost of lost production would have to be more than entirely replacing equipment and storing the spares.

Because safety and protective systems are normally inactive, you may not be able to monitor for deterioration. If the failure is also random, it may not make sense to replace the component on a timed basis because the new part might fail as soon as it has been installed. We simply can't tell—again, because the equipment won't reveal the failure until switched to its active mode. In these cases, some sort of testing may be possible. In our earlier classification of failures we sorted these hidden ones from the rest. If condition monitoring and usage- or time-based intervention aren't practical, you can use RCM logic to explore functional failure-finding tests. These tests can activate the device and reveal whether or not it's working. If such a test isn't possible, redesign the component or system to eliminate the hidden failure. Otherwise, there could be severe safety or protective consequences, which obviously are unacceptable.

For non-hidden problems that can't be prevented through either early prediction or usage- or time-based replacements, you can either redesign or accept the consequences. For safety or environmental cases, where the consequences are unacceptable, redesign is the best decision. For production-related cases, whether to redesign or run to failure may depend on the cost of the consequences. If it is likely that production will completely shut down for a long time, it would be wise to redesign. If the production loss is negligible, run to failure is appropriate. If there aren't any production consequences but there are maintenance costs, the same applies. In these cases, the decision is based on economics—the cost of redesign vs. the cost of failure (e.g., lost production, repair costs, and/or overtime).

Task frequency is often difficult to determine. It is invaluable when using RCM to know the failure history, but, unfortunately, it isn't always available. There isn't any operating history at all, for instance, for a system still being designed. You're up against the same problem with older systems for which records haven't been kept. An option is to use generic failure rates from commercial or private databases. Failures, though, don't happen exactly when predicted. Some will be random, some will become more frequent late in the life of the equipment, and so on. Allow some leeway. Recognize also that the database information may be faulty or incomplete. Be cautious and thoroughly research and deal with each failure mode in its own right.

Once RCM has been completed, you need to group similar tasks and frequencies so that the maintenance plan applies in an actual working environment. You can use a slotting technique to simplify the job. This consists of a predetermined set of frequencies, such as daily, weekly, monthly, every shift, quarterly, semiannually, and annually, or by units produced, distances traveled, or number of operating cycles. Choose frequency slots closest to those that maintenance and operating history shows are best suited to each failure mode. Later, these slots can be used to group tasks with similar characteristics in a workable maintenance plan.

After you run the failure modes through the above logic, consolidate the tasks into a Maintenance Requirements Document. This is the final "product" of RCM. Then, those maintaining and operating the system must continually strive to improve the product. The original task frequencies may be overly conservative or too long. If too many preventable failures occur, it shows that proactive maintenance isn't frequent enough. If there are only small failures or the preventive costs are higher than before, maintenance frequencies may be too high. This is where optimizing task frequency is important.

The Maintenance Requirements Document describes the condition monitoring, time- or usage-based intervention, failure-finding tasks, and the redesign and run-to-failure decisions. It is not a plan in the true sense. It doesn't contain typical maintenance planning information such as task duration, tools and test equipment, parts and materials requirements, trades requirements, and a detailed sequence of procedures. These are logical next steps after RCM has been performed.

The Maintenance Requirements Document is also not merely a series of lists transferred from the RCM worksheets or database into a report. RCM takes you through a rigorous process that identifies and addresses individual failure modes:

- For each plant, there are numerous systems.
- For each system, there can be numerous equipment.
- For each piece of equipment, there are several functions.
- For each function, there may be several failures.

- For each failure, there may be many failure modes with varying effects and consequences.
- For each failure mode, there is a task.

In a complex system, there can be thousands of tasks. To get a feel for the size of the output, consider a typical process plant that has spares for only about 50% of its components. Each may have several failure modes. The plant probably carries some 15,000 to 20,000 individual part numbers (stock-keeping units) in its inventory. That means there could be around 40,000 parts with one or more failure modes. If we analyzed the entire plant, the Maintenance Requirements Document would be huge.

Fortunately, a limited number of condition-monitoring techniques are available—just over 100. These cover most random failures, especially in complex mechanical, electrical, and hydraulic systems. Tasks can be grouped by technique (e.g., vibration analysis), location (e.g., the machine room), and sublocation on a route. Hundreds of individual failure modes can be organized this way, reducing the number of output tasks for detailed maintenance planning. It's important to watch the specified frequencies of the grouped tasks. Often, we don't precisely determine frequencies, at least initially. It may make sense to include tasks with frequencies like "every four days" or "weekly" but not "monthly" or "semiannually." This is where the slotting technique described earlier comes in handy.

Time- or usage-based tasks are also easy to group together. You can assemble all the replacement or refurbishment tasks for a single piece of equipment by frequency into a single overhaul task. Similarly, multiple overhauls in a single area of a plant may be grouped into one shutdown plan.

Another way to organize the outputs is by who does them. Tasks assigned to operators are often performed using the senses of touch, sight, smell, or sound. These are often grouped logically into daily, shift, or inspection rounds checklists. In the end there should be a complete listing that tells what maintenance to do and when. The planner determines what is needed to execute the work.

7.6 WHAT DOES IT TAKE TO DO RCM?

RCM very thoroughly examines plants and equipment. It involves detailed knowledge of how equipment and systems operate, what's in the equipment, how it can fail to work, and the impact on the process and the plant and its environment.

You must practice RCM frequently to become proficient and make the most of its benefits. To implement RCM:

- Select a team of practitioners.
- Train them in RCM.
- Teach other "stakeholders" in plant operations and maintenance what RCM is and what it can achieve.
- Select a pilot project to demonstrate success and improve on the team's proficiency.
- Roll out the process to other areas of the plant.

You must also be able to demonstrate RCM's success. Before the RCM team begins the analysis, determine the plant baseline reliability and availability measures, as well as proactive maintenance program coverage and compliance. These measures will be used later to compare what has been changed and how successfully.

7.6.1 The Team, Skills, Knowledge, and Other Resources Needed

A multidisciplinary team is essential, with specialists brought in when needed. The team needs to know the day-to-day operations of the plant and equipment, along with detailed knowledge of the equipment itself. This dictates at least one operator and one maintainer. They must be hands-on and practical, willing and able to learn the RCM process, and motivated to make it a success. The team must also be versed in plant operations, usually supplied by a senior operations person, such as a supervisor who has risen through the ranks. The team needs to know planning, scheduling, and overall maintenance operations and capabilities to ensure that the tasks are truly doable in the plant. You may have to contract out some of the work to qualified service providers, especially infrequent yet critical equipment monitoring. This expertise comes from someone with a supervisory maintenance background.

Finally, detailed equipment design knowledge is important. The maintainers will know how and why the equipment is put together, but they may have difficulty quantifying the reasons or fully understanding the engineering principles utilized. An engineer or senior technician/technologist from maintenance or production, usually with a strong mechanical or electrical background, is needed on the team.

RCM is very much a learning process for its practitioners. Five team members is about optimal to fulfill the requirements. Too many people slow the progress. Not enough means too much time will be spent trying to understand the systems and equipment.

The team will need help to get started. Someone in house may already have done RCM, but more likely you will have to look outside the company. Training is usually followed up with a pilot project, producing a real product.

RCM is thorough, and that means it can be time-consuming. Training the team usually takes about one week, but can last as long as a month, depending on the approach. Training of other stakeholders can take from a couple of hours to a day or two, depending on their degree of interest and need to know. Senior executives and plant managers should be involved, so they know what to expect and what support is needed. Operations and maintenance management must understand the time demands on their staff and what to expect in return. Finally, operators and maintainers must also be informed and involved. Since their coworkers are on the RCM team, they are probably shouldering additional work.

The pilot project time can vary widely, depending on the complexity of the equipment or system selected for analysis. A good rule of thumb is to allow the team a month for pilot analysis to ensure that they learn RCM thoroughly and are comfortable using it. On average, each failure mode takes about half an hour to analyze. Determining the functional failure to task frequency can take from six to 10 minutes per failure mode. Using our previous process plant example, a very thorough analysis of all systems comprising at least 40,000 items (many with more than one failure mode) would entail over 20,000 man-hours (that's nearly 10 person-years for an entire plant). When divided by five team members, the analysis could take up to two years. It's a big job.

7.7 IS RCM AFFORDABLE?

7.7.1 What to Expect to Pay for Training, Software, Consulting Support, and Staff Time

In the process plant example, RCM requires considerable effort. That effort comes at a price. Ten person-years at an average of, say, $70,000 per person adds up to $700,000 for staff time alone. The training for the team and others will require a couple of weeks from a third party, at consultant rates. The consultant should also be retained for the entire pilot project—that's another month. Even though consulting rates are steep, running into thousands of dollars per day, it's worth it. The price is small when you consider that lives can be saved and environmental catastrophe or major production outages avoided. Numerous experts in the field have seen RCM prevent major safety and production calamities and save considerable maintenance workforce effort.

Software is available to help manipulate the vast amount of analysis data you generate and record. There are several databases to step you through the RCM process and store results. Some of the software costs only a few thousand dollars for a single-user license. Some of it is specifically RCM software while some comes as part of large computerized maintenance management systems. Prices for these high-end systems that include RCM are typically hundreds of thousands of dollars. If you're working with RCM consultants, you'll find that many have their preferred software tools. A word of caution about RCM systems: ensure that they comply with the SAE standard.

To decide on task frequencies, you must know the plant's failure history, which is generally available through the maintenance management system. If not, you can obtain failure rates from databases. You may need help to build queries and run reports. There are external reliability databases you can use, but they often are difficult to find and charge a user a license fee.

7.8 RCM VARIETIES

RCM comes in several varieties, depending on the application. These include:

- Aerospace—commercial airlines, described by Nowlan and Heap (2) and in MSG-3 (3)
- Military—various for naval and combat aircraft, described in numerous U.S. Military Standards (4)
- Commercial, as described by Smith (5) and Moubray (6)
- Streamlined versions, some of which don't meet the SAE JA1011 criteria and are no longer called RCM

The aerospace variation differs in two ways from industrial approaches. First, any structural components are thoroughly analyzed for stress, often using finite element modeling techniques. You identify weaknesses in the airframe structure that must be regularly inspected and undergo nondestructive testing. The second difference is that, once you decide on a maintenance plan, you continue to follow the RCM logic questions on the chance that you may identify a redundant maintenance action. For example, you may conclude that condition-monitoring techniques will identify a particular failure mode, enter that check into your maintenance plan, and then go on to perform time-based replacement. This is a conservative approach, justified by serious safety concerns if an aircraft fails.

The military standards describe the same processes, with examples from military applications, using military equipment terminology.

The commercial versions are what we have described in this book.

7.9 STREAMLINED RCM AND ALTERNATIVE TECHNIQUES

7.9.1 Classical RCM

Classical RCM is a term sometimes used to describe the original process, laid out by Nowlan and Heap, Smith, and Moubray, and referenced in the SAE standard. Classical RCM has proven highly successful in numerous industries, particularly at:

- Reducing overall maintenance effort and costs
- Improving system and equipment performance to achieve design reliability
- Eliminating planned to-be-installed redundancy and reducing capital investment

Many cost and effort reductions have occurred in industries that were:

- Overmaintaining (e.g., civil and military aircraft, naval ships, nuclear power plants)
- Not maintaining, with low reliability (e.g., thermal power plants, haul trucks in mining, water utilities)
- Overly conservative in design practices (e.g., former

public utilities that must now survive in deregulated environments, oil and gas/petrochemical)

Despite these successes, many companies fail in their attempts to implement RCM. The reasons for failure include:

- Lack of management support and leadership.
- Lack of vision about what RCM can accomplish (the RCM team and the rest of the plant don't really know what it's for and what it will do for them).
- No clearly stated reason for doing RCM (it becomes another "program of the month").
- Not enough resources to run the program, especially in "lean manufacturing."
- A clash between RCM's proactive approach and a traditional, highly reactive plant culture (e.g., RCM team members find themselves being pulled from their work to react to day-to-day crises).
- Giving up before RCM is completed.
- Continued errors in the process and results that don't stand up to practical "sanity checks" by rigorous maintainers. This is often due to a lack of full understanding of FMEA, criticality, RCM logic, Condition Monitoring (CM) techniques, which result in Condition-Based Maintenance (CBM) and Time-Based Maintenance (TBM), and reluctance to accept run-to-failure conclusions.
- Lack of available information about the equipment/ systems being analyzed, which isn't necessarily significant but often stops people cold.
- Criticism that RCM-generated tasks seem to be the same as those already in long use in the PM program. It can be seen as a big exercise that is merely proving what is already being done.
- Lack of measurable success early in the RCM program. This is usually because the team hasn't estab-

lished a starting set of measures, an overall goal, and ongoing monitoring.

- Results don't happen quickly enough, even if measures are used. The impact of doing the right type of preventive maintenance (PM) often isn't immediately apparent. Typically, results are seen in 12 to 18 months.
- There is no compelling reason to maintain the momentum or even to start the program (e.g., no legislated requirement, the plant is running well and the company is making money in spite of itself).
- The program runs out of funding.
- The organization lacks the ability to implement RCM results (e.g.: no system that can trigger PM work orders on a predetermined basis).

The above list is a blueprint for failure. One criticism of RCM is that it's the "$1 million solution to the $100,000 problem." This complaint is unfounded. It's how you manage RCM that's usually at fault, not the process itself. All the reasons for failure described above can be traced back to management flaws.

There are many solutions to the problems we've outlined. One that works well is using an outside consultant. A knowledgeable facilitator can help get you through the process and maintain momentum. Often, however, companies stop pursuing change as soon as the consultant leaves. The best chance of success is gained by using help from outside in the early stages of your RCM program. Successful consultants recognize the causes of failure and avoid them.

Where appropriate, methods that shortcut the RCM process can be effective. But make sure they're proper and responsible, especially when there are health or safety risks. All risks should be quantified and managed. (Refer to Chapter 6 for a thorough discussion of risk management.) No matter how effective, shortcuts cannot be considered RCM unless they comply with SAE JA1011.

7.9.2 Streamlined RCM

Some of these shortcut RCM methods have become known as "streamlined" or "lite" RCM. In one variation, RCM logic is used to test the validity of an existing PM program. This approach, though, doesn't recognize what may already be missing from the program, which is the reason for doing RCM in the first place. This is not RCM.

For example, if the current PM program extensively uses vibration and thermographic analysis but nothing else, it probably works well at identifying problems causing vibrations or heat, but not failures such as cracks, fluid reduction, wear, lubricant property degradation, wear metal deposition, surface finish, and dimensional deterioration. Clearly, this program does not cover all possibilities. Applying RCM logic will result, at best, in minor changes to what exists. The benefits may be reduced PM effort and cost, but anything that isn't already covered will be missed. This streamlined approach adds minimal value. In fact, it's irresponsible.

7.9.3 Criticality

Criticality, another RCM variation, is used to weed out failure modes from ever being analyzed. This must be done carefully. One approach is described thoroughly in MIL STD 1629A (7), but there are several different techniques. Basically, failures are not analyzed if their effects are considered noncritical, they occur in noncritical parts or equipment, or they don't exceed the set criticality hurdle rate. Reducing the need for analysis can produce substantial savings.

When criticality is applied to weed out failure modes, there should be relatively little risk of causing a critical problem. Criticality of failure modes can be determined only after the failure modes have been identified—late in the process. Because most of the RCM analysis has already been performed, using criticality is relatively risk-free. The drawback, though, is the effort and cost expended deciding to do nothing. In the end, there are very little savings.

You can cut costs in both areas by reducing RCM analysis before most of it is done. Using a criticality hurdle rate eliminates many possible equipment failures, without having to document them, which can mean big maintenance savings. The downside is that, even when you look at worst-case scenarios, you run the risk that a critical failure will slip through unnoticed. How well this technique works depends on how much the RCM team knows. The greater their plant knowledge and experience, the less the risk. Often, then, the strongest team members are the plant's best maintainers and operators.

This approach may be the right choice if you confidently know and accept the consequences of failure in production, maintenance, cost, and environmental and human terms. For example, many failures in light manufacturing have relatively little fallout other than lost production time. But in other industries, you can be sued if a failure could have been prevented or otherwise mitigated through RCM. The nuclear power, chemical processing, pharmaceutical, automotive, and aircraft industries are especially vulnerable. SAE JA1011 stipulates that the method used to identify failure modes must show what is likely to occur. Of course, the level at which failure modes are identified must be acceptable to the owner or user. There is room for judgment, and, if done properly, this method can meet the SAE criteria that define RCM.

Criticality helps you prioritize so that the most important items are addressed first. It cuts the RCM workload that typically comes with large plants, systems and equipment, and limited resources to analyze them.

Many companies suffer from the failures described, as well as others. Without the force of law, though, RCM standards such as SAE JA-1011 are often treated as mere guidelines that don't need to be followed. Sometimes the people making the decisions aren't familiar with RCM and its benefits. So what do you do if you know that your plant could suffer a failure that can be prevented? You must responsibly and reasonably do whatever it takes to avert a potentially serious situation. Recognize the doors that are open to you and use them to get started.

Similarly, if you can foresee that an RCM implementation is likely to fail, you must eliminate the reason. Even if this doesn't always seem practical or easy, consider the following consequences of not taking action.

Hypothetically, if a company ignores known failures and does nothing, it could be sued, get a lot of negative publicity, and suffer heavy financial loss. The gas tank problem in the early-model Ford Pinto, dramatized in the 1991 movie "Class Action" starring Gene Hackman, is a case in which the court assigned significant damages. In November 1996, a New Jersey court certified a similar nationwide class action lawsuit against General Motors due to rear brake corrosion.

In another hypothetical situation, if you acknowledge a potential problem but discount it as being of negligible risk, and then subsequently experience the failure, you could be blamed for ignoring what was clearly recognized as a risk— sort of damned if you do and damned if you don't!

Risk can never be eliminated entirely, but it can be lessened. Even if you can't fully implement RCM, take at least some positive action and reduce risk as much as possible. Simply reviewing an existing PM program using RCM logic will accomplish very little. You'll gain more, where it counts most, by analyzing critical equipment. If you follow that by moving down the criticality scale, you'll gain even more and eventually successfully complete RCM.

If performing RCM is simply too much for your company, consider an alternative approach that you can achieve. You'll at least reduce risk somewhat. If you do nothing at all, you could be branded as irresponsible later on, which can be deadly for both your business and your professional reputation.

Your ultimate challenge is to convince the decision-makers that RCM is their best course. You need to build credibility by demonstrating clearly that RCM works. Often, however, maintenance practitioners have relatively little influence and control. A maintenance superintendent may be encouraged to be proactive as long as he or she doesn't ask the operators or production staff for help or need upper management's approval for additional funding. Sound familiar?

RCM needs operator and production help, though, to succeed. Without this support, the best attempts to implement an RCM program can flounder.

What can a maintainer realistically do to demonstrate success and increase influence? Realize first of all that many companies do value at least some degree of proactive maintenance. Even the most reactive may do some sort of PM. They know that an ounce of prevention is worth a pound of cure, even if they're not using it to their best advantage. The climate may already exist for you to present your case.

The cost of using RCM logic can be a stumbling block. The bulk of the work—identifying failure modes—is where most of the analysis money is spent. Reduce the cost and you'll generate more interest.

7.9.4 Capability-Driven RCM

Even if you're under severe spending constraints, you can still make improvements by being proactive. By using what's known as Capability-Driven RCM (CD-RCM), you:

- Reverse the logic of the RCM process, starting with the solutions (of which there are a finite number), and look for appropriate places to apply them. Since RCM progresses from equipment to failure modes, through decision making to a result, the opposite process can pinpoint failures, even if they aren't clearly identified.
- Extend existing condition-monitoring techniques to other pieces of equipment. For instance, vibration analysis that works on some equipment can also work elsewhere.

- Look specifically for wearout failures and simply do time-based replacements.
- Check standby equipment to ensure that it works when needed.

These are examples of proactive maintenance that can make a huge difference. It's crucial, though, to do root-cause analysis when a failure does occur, so you can take preventive action for the future. Among the benefits of using CD-RCM is that it can be a means of building up to full RCM.

There are some risks involved in CD-RCM, however. Some items may be overmaintained, especially in cases where run-to-failure has previously been acceptable. Overmaintenance can even cause failures if it disrupts operations and equipment in the process. The risk, though, is relatively small. The bigger problem may be the cost of using maintenance resources that aren't needed.

There is also a risk of missing some failure modes, maintenance actions, and redesign opportunities that could have been predicted or prevented if upfront analysis had been done. While the consequences can be significant, the risk is usually minimal if the techniques used are broad enough. For instance, Nowlan and Heap (2) found that condition monitoring is effective for airlines because approximately 89% of aircraft failures in their study were not time-related.

CD-RCM can be used in a similar way with other complex electromechanical systems using complicated controls and many moving parts. In industrial plants, for instance, looking for wearout failure modes is like the traditional approach to maintenance. Most failures are influenced by operating time or some other measure. By searching for failure conditions only where parts are in moving contact with each other and with the process materials, there will be fewer items to be examined. The potential failures are generally obvious and easy to spot. With this approach it is possible to overmaintain, especially if the equipment you decide to perform TBM on has many other random failure modes.

In RCM, failure finding is used for hidden failures that either are not detected using CBM or avoided using TBM. One favored method is to run items that are usually "normally off" to test their operation. This is done under controlled conditions so that a failure can be detected without significant problems occurring. There is a failure risk on startup in activating the system or equipment, but there is also control over the consequences because the check is done when the item isn't really needed. Correcting the failure reduces its consequences during "normal" operation, when it would be really needed if the primary equipment or device failed. Doing failure-finding tasks without knowing what you're looking for may seem foolish, but it's not. Failures often become evident when the item is operated outside of its normal mode (which is often "off"). Although all "hidden failures" may not be found, many will.

Again, this is not as thorough as a complete RCM analysis, but it's a start in the right direction. By showing successful results and thus gaining credibility, the maintainer may be able to extend his proactive approach to include RCM analysis. CD-RCM is not intended to avoid or shortcut RCM. It is a preliminary step that provides positive results consistent with RCM and its objectives.

To be successful, CD-RCM must:

- Ensure that the PM work order system actually works (i.e., PM work orders can be triggered-automatically; the work orders get issued and carried out as scheduled). If this is not in place, help is needed that is beyond the scope of this book.
- Identify the equipment/asset inventory (this is part of the first step in RCM).
- Identify the available conditioning-monitoring techniques that may be used (which is probably limited by plant capabilities).
- Determine the kinds of failures that each of these techniques can reveal.

- Identify the equipment in which these failures dominate.
- Decide how often to monitor and make the process part of the PM work order system.
- Identify which equipment has dominant wearout failure modes,
- Schedule regular replacement of wearing components and others that are disturbed in the process.
- Identify all standby equipment and safety systems (alarms, shutdown systems, stand by and redundant equipment, backup systems, etc.) that are normally inactive but needed in special circumstances.
- Determine appropriate tests that will reveal failures that can be detected only when the equipment runs. The tests are then implemented in the PM work order system.
- Examine failures that are experienced after the maintenance program is put in place to determine their root cause so that appropriate action may be taken to eliminate them or their consequences.

The result of using CD-RCM can be:

- Extensive use of CBM techniques such as vibration analysis, lubricant/oil analysis, thermographic analysis, visual inspections, and some nondestructive testing
- Limited use of time-based replacements and overhauls
- In plants where redundancy is common, extensive "swinging" of operating equipment from A to B and back, possibly combined with equalization of running hours
- Extensive testing of safety systems
- Systematically capturing and analyzing information about failures that occur, to determine the causes and eliminate them in the future

All these actions will move the organization to be more proactive. As CD-RCM targets proven methods where they make sense, it builds credibility and enhances the likelihood of implementing a full SAE-compliant RCM program.

7.10 HOW TO DECIDE?

7.10.1 Summary of Considerations and Tradeoffs

RCM is a lot of work (Figure 7-4). It is also expensive. The results, although impressive, can take time to accomplish. One challenge of promoting a RCM program is justifying the cost without being able to show concretely what the savings will be.

The cost of *not* using RCM, however, may be much higher. Some alternatives to RCM are less rigorous and downright dangerous. We believe that RCM is a thorough and complete approach to proactive maintenance that achieves high system reliability. It addresses safety and environmental concerns,

Figure 7-4 It can be a lot of work.

identifies hidden failures, and results in appropriate failure-finding tasks or checks. It identifies where redesign is appropriate and where run-to-failure is acceptable or even desirable.

Simply reviewing an existing PM program with an RCM approach is not really an option for a responsible manager. Too much can be missed that may be critical, including safety or environmental concerns. It is not RCM.

Streamlined or "lite" RCM may be appropriate for industrial environments where criticality is an issue. RCM results can be achieved on a smaller but well-targeted subset of the failure modes on the critical equipment and systems. Although this is a form of RCM applied to a subset of the plant, care must be taken to ensure that it meets the SAE criteria for RCM.

Where RCM investment is not an immediately achievable option, the final alternative is to build up to it using CD-RCM. This adds a bit of logic to the old approach of applying a new technology everywhere. In CD-RCM, you take stock of what you can do now and make sure you use it as widely as possible. Once success has been demonstrated, you can expand on the program. Eventually, RCM can be used to make the program complete.

7.10.2 RCM Decision Checklist

Throughout this chapter, you have had to consider certain questions and evaluate alternatives to determine if RCM is needed. We summarize them here for quick reference.

1. Can the plant or operation sell everything it can produce? If the answer is "yes," high reliability is important and RCM should be considered. Skip to question 5. If the answer is "no," focus on cost-cutting measures.
2. Does the plant experience unacceptable safety or environmental performance? If "yes," RCM is prob-

ably needed. Skip to question 5. Is an extensive preventive-maintenance program already in place? If the answer is "yes," consider RCM if the PM program costs are unacceptably high. If "no," consider RCM if total maintenance costs are high compared with others in the same business.

3. Are maintenance costs high relative to others in your business? If "yes," RCM is right. Proceed to question 5. If not, RCM won't help. Stop here.

At this point, one or several of the following apply:

- A need for high reliability
- Safety or environmental problems
- An expensive and low-performing PM program
- No significant PM program and high overall maintenance costs

4. RCM is right for your organization. Next, you need to ensure that the organization is ready for it. Is there a "controlled" maintenance environment where most work is predictable and planned? Does planned work, like PM and predictive maintenance (PdM), generally get done when scheduled? If "yes," the organization passes this basic test of readiness—the maintenance environment is under control. Proceed to question 6. RCM won't work well if it can't be applied in a controlled environment. If this is the case, RCM alone isn't enough. Get the maintenance activities under control first. Stop here.

5. RCM is needed and the organization is ready for it. But senior management support is probably still required for the investment of time and cost in RCM training, piloting, and rollout. If that is not forthcoming, consider the alternatives to full RCM. Proceed to question 7.

6. Can senior management support be obtained for the investment of time and cost in RCM training and piloting? This investment will require about one month of team time (five people) plus a consultant for the month. If "yes," consider RCM pilot to demonstrate success before attempting to roll out RCM across the entire organization.
7. If "no," you must prove credibility to senior management with a less thorough approach that requires little upfront investment and uses existing capabilities. The remaining alternative here is CD-RCM and a gradual buildup of success and credibility to expand on it.

REFERENCES

1. SAE JA1011. Evaluation Criteria for Reliability-Centered Maintenance (RCM) Processes. Society of Automotive Engineers, Aug 1999.
2. F. S. Nowlan, H. Heap. Reliability-Centered Maintenance. Report AD/A066-579. National Technical Information Service, Dec 19, 1978.
3. MSG-3. Maintenance Program Development Document. Revision 2. Washington, DC: Air Transport Association, 1993.
4. MIL-STD 2173 (AS), Reliability-Centered Maintenance Requirements for Naval Aircraft, Weapons Systems and Support Equipment, US Naval Air Systems Command; NAVAIR 00-25-403, Guidelines for the Naval Aviation Reliability Centered Maintenance Process, US Naval Air Systems Command; S9081-AB-GIB-010/MAINT—Reliability-Centered Maintenance Handbook, US Naval Sea Systems Command.
5. Anthony M. Smith. Reliability Centered Maintenance. New York: McGraw-Hill, 1993.
6. John Moubray. Reliability-Centered Maintenance. 2nd ed. Oxford: Butterworth-Heinemann, 1997.
7. MIL STD 1629A, Notice 2. Procedures for Performing a Failure Mode, Effects and Criticality Analysis. Washington, DC: Department of Defense, 1984.

8

Reliability by Operator: Total Productive Maintenance

Total Productive Maintenance (TPM) is a highly powerful philosophy for managing maintenance, operations, and engineering in a plant environment. It harnesses the power of the entire workforce to increase the productivity of the company's physical assets, optimizing human–machine interaction. It is an internal continuous-improvement process to meet increasingly difficult market demands and to provide mass customization for individual customers. Only with a highly flexible manufacturing process and workforce can a company achieve this.

In this chapter you will learn the fundamental functions of TPM, what they mean and how they are used and integrated into a comprehensive program. When completely implemented, TPM becomes more than a program to run the plant—it becomes part of the culture.

We also explore the implementation issues that you can expect and compare a TPM approach with typical legacy environments. This will dispel some of the many myths about TPM.

Finally, we link TPM to other optimizing methodologies discussed elsewhere in this book. With their combined effect, you can implement a true continuous-improvement environment.

8.1   INTRODUCTION: WHAT IS TPM?

TPM is the most recognized continuous-improvement philosophy, but it's also the most misunderstood. It has the power to radically change your organization and boost overall production performance.

Some people claim that TPM reflects a certain culture and isn't applicable everywhere. That's been proven wrong countless times. Others maintain that TPM is just common sense, but there are plenty of people with common sense who haven't been successful using TPM. Clearly, TPM is much more than this.

The objectives of TPM are to optimize the relationship between human and machine systems and the quality of the working environment. What confuses skeptics is the approach that TPM uses to eliminate the root causes of waste in these areas.

TPM recognizes that the roles of Engineering, Operations, and Maintenance are inseparable and codependent. It uses their combined skills to restore deteriorating equipment, maintain basic equipment and operating standards, improve design weaknesses, and prevent human errors. The old paradigm of "I break, you fix" is replaced with "together we succeed." This is a radical change for many manufacturing and process organizations.

TPM is more about changing your workplace culture than adopting new maintenance techniques. For this reason, it can be agonizingly difficult to implement TPM even though its concepts seem so simple on the surface. As a result, the published TPM methodologies are associated with implementation techniques. Technical change is rapid, but social change takes time and perseverance. Most cultures need an external stimulus. Modern manufacturing philosophies, specifically Just-In-Time (JIT) and Total Quality Management (TQM), are market driven—they force an organization to make a cultural change. TPM has grown along with the need for flexible manufacturing that can produce a range of products to meet highly variable customer demands. Once TPM is in place it continues to develop and grow, promoting continuous improvement. A TPM organization drives change from within.

In companies that have developed a thorough understanding of TPM, it stands for Total Productive Manufacturing. This recognizes that TPM encompasses more than maintenance concerns, with the common goal of eliminating all waste in manufacturing processes. TPM creates an orderly environment in which routines and standards are methodically applied. Combining teamwork, individual participation, and problem-solving tools maximizes your equipment use.

What do you need to develop a TPM culture? Besides the tools of TPM, it requires production work methods, production involvement in minor maintenance activities, and teaming production and maintenance workers. The operator becomes key to machine reliability, rather than a major impediment as many believe.

This concept must be accepted and applied at all levels of your organization, starting from the bottom up, and nurtured by top management. The result is an organization committed to the continuous improvement of its working environment and its human–machine interface.

8.2 WHAT ARE THE FUNDAMENTALS OF TPM?

TPM has five fundamental functions:

1. Autonomous maintenance
2. Equipment improvement
3. Quality maintenance
4. Maintenance prevention
5. Education and training

8.2.1 Autonomous Maintenance

Many people confuse autonomous maintenance (AM) with TPM, but it is only one part of TPM, although a fundamental one. The confusion arises because during a TPM implementation AM directly affects the most people.

AM is a technique to get production workers involved in equipment care, working with Maintenance to stabilize conditions and stop accelerated deterioration. You must teach operators about equipment function and failures, including prevention through early detection and treating abnormal conditions.

AM can create conflict because of past work rules. For it to be successful, operators must see improvements, strong leadership, and control elements delivering satisfactory service levels. See Campbell (1) for a thorough understanding of maintenance leadership and control—the platform for developing TPM.

AM's impact on maintenance is often overlooked. In fact, it helps your staff support the operators, make improvements, and solve problems. More time is spent on maintenance diagnostics, prevention, and complex issues. Operators perform routine equipment inspections and CLAIR—cleaning, lubrication, adjustment, inspection, and (minor) repair—maintenance tasks, which are critical to how the equipment performs.

The first AM task is to have equipment maintainers and operators complete an initial cleaning and equipment rehabilitation. During this time the operators learn the details of their equipment and identify improvement opportunities. They learn that "cleaning is inspection," as described by the Japan Institute of Plant Maintenance (2). Regular cleaning exposes hidden defects that affect equipment performance. Inspection routines and equipment standards are established. The net effect is that the operator becomes an expert on his or her equipment.

This development of the operator as an expert is critical to the success of TPM. An expert operator can judge abnormal from normal machine conditions and communicate the problem effectively, while performing routine maintenance. It is precisely this expert care that maximizes equipment.

8.2.2 Equipment Improvement

A key function of TPM is to focus the organization on a common goal. Since people behave the way they are measured, it is critical to develop a comprehensive performance measure for all employees. The key TPM performance measure is Overall Equipment Effectiveness (OEE), as described by Nakajima (3).

OEE combines equipment, process, material, and people concerns and helps identify where the most waste occurs. It focuses Maintenance, Engineering, and Production on the key issue of plant output.

Using OEE, you will be able to better identify whether your operation is producing quality product. Simply put, an operation either is always producing on-spec product or it is not. OEE forces the organization to address all the reasons for lost production, turning losses into opportunities for improvement.

As shown in Figure 8-1, OEE is the measurement of all equipment activities in a given period. At any given time, equipment will always perform one of the following: on-spec product, downtime, quality loss, or rate loss. The size of the total pie in Figure 8-1 is the amount of product that could be

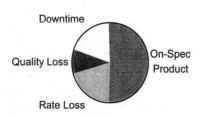

Figure 8-1 OEE: time allocation of activities.

produced at the ideal rate for a period of calendar time. This is the ideal state where all of the organization's efforts produce on-spec product.

Often, an organization will decide to remove calendar time from the OEE calculation. If your plant operates 5 days a week, you may want to eliminate the downtime caused by not operating on Saturday and Sunday, but keep in mind that this OEE calculation is really a subset of the OEE for the plant. In Figure 8.1, a decision has been made to reduce the size of the pie.

OEE is calculated by the formula

$$OEE = Availability\ \% \times Production\ rate\ \% \\ \times Quality\ rate\ \%$$

where:

$$Availability = \frac{Production\ time}{Total\ time}$$

$$Production\ rate = \frac{Actual\ production\ rate}{Ideal\ production\ rate}$$

$$Quality\ rate = \frac{Actual\ on\text{-}spec\ production}{Actual\ total\ production}$$

Availability is simply the ratio of production time to calendar time. In practice, it is more convenient to measure the downtime and perform some simple mathematics to arrive at the production time. Downtime is any time the equipment is not producing. Equipment, systems, or plants may be shut down but available for production.

The downtime could be unrelated to the equipment—caused, for example, by a lack of raw materials to process. Count all downtime, including any you have scheduled. Excluding some downtime violates the key principle that no one should "play" with the numbers.

At one plant, planned maintenance was excluded from the availability calculation. If an 8-hour shift had 2 hours of planned maintenance, the total time for that shift was said to be 6 hours. Supervisors reacted by calling almost any maintenance activity "planned." While OEE went up, total output did not. Downtime is downtime, planned or not.

Production rate is the actual product ratio when running at the "ideal" instantaneous production rate. Setting the ideal rate can be difficult because there are different approaches to determining its value. The most common approach is to use either a rate shown to be achievable or the design production rate, whichever is greater. The design rate is a good target if your plant, system, or equipment has never achieved it—common in newer plants. If the plant has been modified over the years and production capacity has increased, or you've used various "tricks" to increase production levels successfully, the demonstrated maximum production rate is useful. In many cases, this corresponds to an upper control limit.

You can measure production rate for continuous operation "on the fly" by just looking at the production speed indicators. Batch processes, however, can be difficult to measure. In a batch process, cycle time, or average rate based on production time output, is used to measure production rate.

Quality rate is the ratio of on-spec product to actual production rate. On-spec means producing what is needed, in a condition that complies with product specifications. Product that does not meet spec may be salable to some lower spec, if customers order large enough quantities.

An often forgotten part of the quality rate is rework, which shouldn't affect the production rate. Where rework is fed back through the process (for example, steel scrap from the hot strip mill is fed back to the basic oxygen furnace at a steel mill), it displaces virgin material that could be processed in its place.

Keep in mind that knowing the OEE doesn't provide information to improve it. You need to determine what causes each loss and how significant it is. If you know where most waste occurs, you can focus resources to eliminate it through problem solving, RCFA (see Chapter 2), RCM (see Chapter 7), and some very basic equipment-care techniques.

8.2.3 Quality Maintenance

You have implemented OEE in your key production areas, and you have a wealth of data about your losses. But how do you use that data to improve the OEE?

Although the OEE number is the focus, identifying what causes wasted availability and process rate is essential to improving it, along with applying quality Pareto analysis. By working on the most significant losses, you make the most effective use of your resources. There are many forms of waste that TPM can eliminate:

- Lost production time due to breakdowns
- Idling and minor stoppage losses from intermittent material flows
- Setup and adjustment losses (time lost between batches or product runs)
- Capacity losses from operating the process at less than maximum sustainable rates
- Startup losses from running up slowly or disruptions
- Operating losses through errors
- Yield losses through less than adequate manufacturing processes
- Defects in the products (quality problems)
- Recycling losses to correct quality problems

You can apply OEE at the plant, production line, system, work-cell, or equipment level. It can be measured yearly, monthly, weekly, daily, by shift, by hour, by minute, or instantaneously. The measurement frequency must ensure that both random and systematic events are identified. You must report the data frequently enough to detect trends early. A 90% OEE target is world-class. To get there, first improve availability (largely a maintenance and reliability effort), then target production and quality rates.

However, as you can see in Figure 8-2, using Pareto analysis, as described by Ishikawa (4), is the key to improving OEE. Losses caused by operating at a less than ideal rate or producing off-spec product can be converted to time. That is, a machine operating at 90% of rated speed for 10 hours has lost the equivalent of 1 hour of production time. Pareto analy-

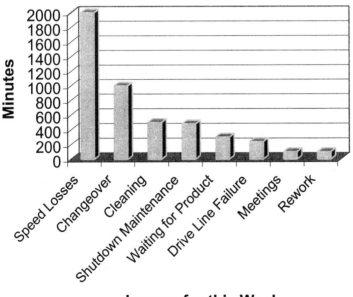

Losses for this Week

Figure 8-2 Sample Pareto analysis.

sis prioritizes the losses so that the organization focuses on the largest piece.

In most organizations, there is a narrow set of measures that zero in on defects or failures. Many organizations monitor mechanical downtime but not availability. In Figure 8-2, the organization would try to correct the drive-line failure. However, completely removing the cause of drive-line failure would be equivalent to a 10% reduction in the amount of product lost to operating at a reduced speed.

Pareto analysis is critical to prioritizing OEE data. Note that the five highest causes in this example are losses often considered normal. In a non-TPM plant drive-line failure and rework would receive the most management attention.

The solution to many of these losses extends beyond maintenance to include production, engineering, and materials logistics. All elements of the plant's entire supply chain can impact how much quality product is produced.

Correcting problems that lead to low availability, production rates, or quality rates can involve maintenance, engineering, and production process or procedural changes. Teamwork is essential to pulling these disciplines together.

Key to TPM is the use of teams. Usually you organize the teams around production areas, lines, or work cells. They comprise production and maintenance workers in a ratio of about 2 to 1. The teams work mostly in their assigned areas to increase equipment familiarity, a sense of ownership, and cooperation between Production and Maintenance. Selecting the pilot area is important. It must obviously need change. You want impressive results that you can use later to "sell" the concept to other plant areas. Through teaming, production and maintenance goals are the same, because they are specific to the area instead of to a department or function.

8.2.4 Maintenance Prevention

You can eliminate a lot of maintenance by studying equipment weaknesses. There are many maintenance-prevention techniques, from simple visual controls to relatively complex engineering design improvements, that can greatly reduce losses. This is very important for effective autonomous maintenance, such as reducing cleaning requirements or increasing ease of adjustment.

Using maintenance prevention effectively converts random breakdown maintenance to routine scheduled maintenance. When operators use lubricants properly, they prevent many unnecessary failures. As the scope of the preventive maintenance program and availability of maintenance tactics increase, problems will be found before failure occurs.

A valuable maintenance prevention tool is an effective Computerized Maintenance Management System (CMMS) or Enterprise Asset Management (EAM) system. The CMMS data helps you prevent recurring failures and effectively plan and schedule maintenance tasks.

8.2.5 Education and Training

Individual productivity is a function of skills, knowledge, and attitude. The drastic change in the operator's work environment when implementing TPM makes education and training essential.

In many organizations, operators are supposed to follow the supervisor's directions without question. The operator is trained never to deviate from a specific procedure. Often, the training method is the "buddy system." The result, however, can be operators who complete the minimum work required to perform the task without any understanding of their role in the overall operation. The operator may also learn bad habits from his or her "buddy."

In TPM, the operator is asked to participate in the decision-making process and constantly question the status quo. This is a basic requirement of continuous improvement. The initial impact, however, can be negative. The operator's first reaction is often that management is "dumping" work that has traditionally been the responsibility of maintenance or plant management. The operator may also worry about not being able to do what's required. It's no wonder that in many plants the mention of TPM immediately mobilizes the union.

You absolutely must educate employees about the benefits of TPM and your business needs while training them to use the tools. Education is about developing an individual into a whole person, while training provides specific skills. But you must implement and update education and training at the same time. The level of each must increase as the operator learns new concepts and skills.

Before starting TPM, the operators need to learn its philosophy and practices. They must also know about their company. If you are to involve operators in decision making, they must understand the context. The minimum training requirements are:

- An introduction to TPM
- General inspection techniques
- Diagnostic techniques
- Analytical problem-solving techniques
- Selected technical training

You need to appoint a special TPM team to teach operators and other personnel specific problem-solving methods such as Pareto analysis, Root Cause Failure Analysis, and statistical process control for them to take a more proactive role in the organization. Training and education must be ongoing, to ensure knowledge transfer and keep skills totally up to date.

8.3 HOW DO YOU IMPLEMENT TPM?

TPM is implemented in four major phases:

1. Establish acceptable equipment operating conditions to stabilize reliability.

2. Lengthen asset life.
3. Optimize asset conditions.
4. Optimize life-cycle cost.

The fundamentals continue and expand as you implement the four phases.

The first phase stabilizes reliability by restoring equipment to its original condition. This is done by cleaning the equipment and correcting any defects. Note major problems and establish a plan to resolve them. Make sure operators get sufficient training to turn simple equipment cleaning into a thorough inspection to spot machine defects.

The second phase maintains the equipment's basic operating condition. Standards to do so are developed. Begin data collection and set equipment condition goals. The operators perform minor maintenance activities to eliminate abnormal wear.

In phase 3, improve the equipment's operation from its stabilized level. Cross-functional teams should target chronic losses to increase overall machine performance. Review and update standards. Find and analyze opportunities to increase equipment performance and operating standards beyond original capabilities.

Phase 4 is about optimizing the cost of the asset over its entire life. You achieve this by extending equipment life, increasing performance, and reducing maintenance cost. Keep the machine at its optimal condition. Regularly review processes that set and maintain operating conditions. New equipment will become part of the TPM process. Operator reliability is "built-in."

A cross-functional team approach should be used during all phases of TPM implementation. Building effective teams is a prerequisite to entrenching TPM ideas and behaviors.

TPM principles and techniques are simple and straightforward. The initial focus will be project management and carefully applying change management. Change is central to

Table 8-1 Implementing TPM: The Legacy Culture

Legacy approach	TPM approach	Change-management issues
Clear lines of responsibility exist between production and maintenance employees. When a machine breaks down, operators call maintenance.	Employees work together to solve problems. It is recognized that production and maintenance are inseparable and problems need to be solved jointly.	Employees may believe that the goal of TPM is to eliminate maintenance jobs. Most people equate productivity gains to job losses. It is difficult to see TPM's objectives.
Supervisors direct employee actions. Employees do what they are told when they are told.	Self-directed teams develop and execute plans to achieve progressive goals.	Front-line supervisors have difficulty changing to their new "coach" role. Many don't trust their employees.
Management announces a new program to improve operations. Employees are trained. This is commonly referred to as the "Flavor of the Month."	Management announces that TPM will be implemented. TPM training is conducted and a TPM team formed. A pilot site is chosen and work begins to improve the condition and performance of the pilot site.	At first, few employees recognize that TPM will actually be implemented. When the TPM team shows progress and starts to make improvements, some employees will see the benefits and want to participate. Others will fear the change and reject the process. Over time the naysayers are converted or fade away. This is an exceptionally difficult issue when a plant has failed at a TPM implementation in the past.

Increased production level is achieved in one of two ways: employees are pushed to work faster or equipment must be added.	Reducing losses due to availability, quality rate, and production rate increases production level. The bigger the loss, the greater the potential benefit.	Operators believe that OEE is implemented to rate them and make them work harder. In fact it is the losses that make them work harder. Eventually they begin to see that OEE helps them quantify problems that they have always wanted corrected.
The relationship between Union and Management is adversarial. Each tries to beat the other in negotiations. Grievances and disciplinary actions are used as negotiating tools.	Union and Management work together to achieve the goals of TPM. Each represents its own interests, but negotiations are considered successful if each side benefits.	TPM is a process that does what most unions have always wanted: it gives employees a voice in their workplace and considers them a valuable resource. However, if the union is not involved from the beginning, its distrust of Management's intentions will be a significant hurdle for the TPM team.
Performance measures exist for each department. Maintenance is evaluated on downtime and production output.	OEE is the key measurement. The organization is evaluated on OEE. Pareto analysis prioritizes losses that affect OEE.	Invariably, when OEE tracking starts, big losses are considered normal—changeovers, for example. It is difficult for people to accept that something they have accepted for years can be changed.

TPM. If your organization isn't used to it, or has a history of unsuccessful changes, this will be a major obstacle to success. We recommend a pilot project to demonstrate success in one area before tackling TPM throughout the entire plant.

The choice of location for a TPM pilot is critical. The pilot area must clearly need improving and be visible to as many people as possible. Once you establish momentum, apply TPM to other areas. Divide it into manageable portions and implement one at a time.

Successful TPM requires a transfer of responsibility between management and employees. It depends on a sincere and dedicated management team that pays appropriate attention to change-management issues. The legacy culture presents the greatest change-management problems. See Table 8-1 for samples of typical change-management issues.

TPM is implemented gradually over what can be several (3 to 5) years. Once established, it becomes part of the plant's way of doing things—its culture. Since TPM is about changing the behavior of both workers and managers, it requires patience and positive reinforcement to achieve permanent change. With the right focus and commitment, even very poorly maintained plants can be "turned around" through TPM.

8.4 THE CONTINUOUS-IMPROVEMENT WORKPLACE

Successfully implementing TPM creates an efficient, flexible, and continually improving organization. The process may be long and arduous but, once TPM has been accepted, it is as hard to remove as the culture it replaced. This is significant because managers may change but TPM will continue.

The TPM workplace is efficient because it follows tested procedures that are continually reviewed and upgraded. Change is handled fluidly because effective education and training prepare the workforce to participate in the decision-making process.

TPM embraces other optimizing maintenance management methodologies. RCM and RCFA are often very effective in a TPM environment. RCM impacts the preventive and predictive maintenance program and RCFA improves specific problem areas. TPM impacts the working environment in virtually all respects—the way Production and Maintenance employees work, are organized, use other techniques such as RCM and RCFA, solve problems, and implement solutions.

Your competitors may be able to purchase the same equipment, but not the TPM experience. The time required to implement TPM makes it a significant competitive advantage, one that can't be easily copied.

REFERENCES

1. John D. Campbell. Uptime: Strategies for Excellence in Maintenance Management. Portland, OR: Productivity Press, 1994.
2. Japan Institute of Plant Maintenance. Autonomous Maintenance for Operators. Portland, OR: Productivity Press, 1997.
3. Seiichi Nakajima. Introduction to TPM: Total Productivity Maintenance. Portland, OR: Productivity Press, 1988.
4. Kaoro Ishikawa. Guide to Quality Control. Asian Productivity Organization, 1986.

9

Reliability Management and Maintenance Optimization: Basic Statistics and Economics

As global industrial competitiveness increases, showing value, particularly in equipment reliability, is an urgent business requirement. Sophisticated, user-friendly software is integrating the supply chain, forcing maintenance to be even more mission-critical. We must respond effectively to incessantly fluctuating market demands. All of this is both empowering and extremely challenging. Mathematical statistical models are an invaluable aide. They help you increase your plant's reliability and efficiency, at the lowest possible cost.

This chapter is about the statistical concepts and tools you need to build an effective reliability management and maintenance optimization program. We'll take you from the basic concepts to developing and applying models for analyzing common maintenance situations. Ultimately, you should know how to determine the best course of action or general policy.

We begin with the relative frequency histogram to discuss the four main reliability related functions: the Probability Density Function, the Cumulative Distribution Function, the Reliability Function, and the Hazard Function. These functions are used in the modeling exercises in this and subsequent chapters. We describe several common failure distributions and what we can learn from them to manage maintenance resources. The most useful of these is the Weibull distribution, and you'll learn how to fit that model to a system or component's failure-history data. Finally, we relate economic considerations, specifically the time value of money, to making maintenance decisions. The result is the concept of equivalent annual cost.

In this chapter, the word *maintenance, repair, renewal,* and *replacement* are used interchangeably. The methods we discuss assume that maintenance will return equipment to "good-as-new" condition.

9.1   INTRODUCTION: THE PROBLEM OF UNCERTAINTY

Faced with uncertainty, our instinctive, human reaction is often fear and indecision. We would prefer to know when and how things happen. In other words, we would like all problems and their solutions to be *deterministic*. Problems in which timing and outcome depend on chance are probabilistic or *stochastic*.

Many problems, of course, fall into the latter category. Our goal is to quantify the uncertainties to increase the success of significant maintenance decisions. The methods described in this chapter will help you deal with uncertainty, but our aim is greater than that. We hope to persuade you to treat it as an ally rather than an unknown foe.

"Failure is the mother of success." "A fall in the pit is a gain in the wit." If your maintenance department uses reliability management, you'll appreciate this folk wisdom. In an enlightened environment, the knowledge gained from failures is converted into productive action. To achieve this requires a sound quantitative approach to maintenance uncertainty. To start, we'll show you an easily understood Relative Frequency Histogram of past failures. Also, we look at the Probability Density Function, the Cumulative Distribution Function, the Reliability Function, and the Hazard Function.

9.2 THE RELATIVE FREQUENCY HISTOGRAM

Assume that 48 items purchased at the beginning of the year all fail by November. List the failures in order of their failure ages. Group them, as in the table in Figure 9-1, into convenient time segments—in this case by month—and plot the number of failures in each segment. The high bars in the center represent the highest (or most probable) failure times: March, April, and May. By adding the number of failures occurring before April—14—and dividing by the total number of items—48—the *cumulative probability* of the item failing

Month	Jan	Feb	Mar	Apr	May	June	July	Aug	Sep	Oct
Failures	2	5	7	8	7	6	5	4	3	1

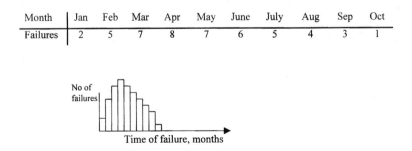

Figure 9-1 The Relative Frequency Histogram.

in the first quarter of the year is 14/48. The probability that all the items will fail before November is 48/48, or 1.

Transforming the numbers of failures by month into probabilities, the Relative Frequency Histogram is converted into a mathematical, and more useful, form called the *probability density function* (PDF). The data is replotted so that the area under the curve, between time 0 and any time *t* represents the cumulative probability of failure. This is shown in Figure 9-2. (How the PDF plot is calculated from the data and then drawn is discussed more thoroughly in Section 9.7.) The total area under the curve of the PDF *f*(*t*) is 1, because sooner or later the item will fail. The probability of the component failing at or before time *t* is equal to the area under the curve between time 0 and time *t*. That area is F(*t*), the Cumulative Distribution Function (CDF). It follows that the remaining (shaded) area is the probability that the component will survive to time *t*. This is known as the reliability function, R(*t*). R(*t*) and F(*t*) can themselves be plotted against time.

The mean time to failure (MTTF) is

$$\int_0^\infty tf(t)dt$$

or the area below the reliability curve (shown in Appendix 1). From the reliability R(*t*) and the Probability Density Function f(*t*), we derive the fourth useful function, the Hazard Function,

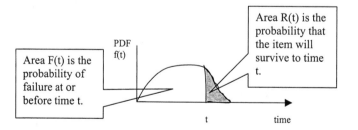

Figure 9-2 The Probability Density Function.

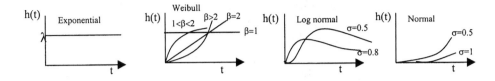

Figure 9-3 Hazard function curves for the common failure distributions.

$h(t) = f(t)/R(t)$, which is represented graphically for four common distributions in Figure 9-3.

In just a few short paragraphs you have discovered the four key functions in reliability engineering. Knowing any one, you can derive the other three. Armed with these fundamental statistical concepts, you can battle random failures throughout your plant. Although we can't predict when failures occur, we can determine the best times for preventive maintenance and the best long-run maintenance policies.

Once you're reasonably confident about the reliability function, you can use it, and its related functions, with optimization models. Models describe typical maintenance situations by representing them as mathematical equations. That makes it convenient to adjust certain decisions to get the optimal outcome. Your optimization objective may be to reduce long-term maintenance costs to the lowest point possible. Other objectives include the highest reliability, maintainability, and availability of operating assets.

9.3 TYPICAL DISTRIBUTIONS

In the previous section we defined the four key functions you may use once data has been transformed into a probability distribution. The prerequisite step of fitting, or modeling, the data is covered next.

How do you find the appropriate reliability function for a real component or system? There are two approaches to this problem. In one, you estimate the reliability function by curve-fitting the failure data from extensive life testing. In the other, you estimate the parameters (unknown constants) by statistical sampling and numerous statistical confidence tests (1). We'll take the latter approach.

Fortunately, we know from past failures that PDFs (and their derived Reliability, Cumulative Distribution, and Hazard Functions) of real maintenance data usually fit one of several mathematical equations already familiar to reliability engineers. These include the exponential, Weibull, log normal, and normal distributions. Each of these represents a "family" of equations (or graphical curves) whose members vary in shape because of their differing parameter values. Their nature can therefore be fully described by knowing the value of their parameters. For example, the Weibull (two-parameter) CDF is:

$$F(t) = 1 - e^{-(t/\eta)^\beta}$$

You can estimate the parameters β and η using the methods described in the following sections. Usually, through one or more of the four functions of Section 9-2, you can conveniently process failure and replacement data. The modeling process involves manipulating the statistical functions you learned about in Section 9.2 (the PDF, CDF, Reliability, and Hazard Functions). The objective is to understand the problem, forecast failures, and analyze risk to make better maintenance decisions. Those decisions will impact the times you choose to replace, repair, or overhaul machinery, as well as optimize many other maintenance management tasks.

The solution entails:

- Collecting good data
- Choosing the appropriate function to represent your situation and estimating the function parameters (for example, the Weibull parameters β and η)
- Evaluating how much confidence you have in the resulting model

Modern reliability software makes this process easy and fun. What's more, it helps us communicate with management and share the common goal of business: implementing procedures and policies that minimize cost and risk while maintaining—even increasing—product quality and throughput.

The most common hazard functions are depicted in Figure 9-3. They correspond to the exponential, Weibull, log normal, and normal distributions. The observed data most frequently fits the Weibull distribution. Today, Weibull analysis is the leading method in the world for fitting component life data (2), but it wasn't always so. While delivering his hallmark paper in 1951, Waloddi Weibull modestly said that his analysis "may sometimes render good service." That was an incredible understatement, but initial reaction varied from disbelief to outright rejection. Then the U.S. Air Force recognized Weibull's research and provided funding for 24 years. Finally, Weibull received the recognition he deserved.

9.4   THE ROLE OF STATISTICS

Most items change into an unacceptable state at some stage in their life, and one of the challenges of optimizing maintenance decisions is to predict when. Luckily, it's possible to analyze previous performance and identify when the transition from "good" to "failed" is likely to occur. For example, while your household lamp may be working today, what is the likelihood that it will work tomorrow? Given historical data on the lifetime of similar lamps in a similar operating environment, you can calculate the probability of the lamp's still working tomorrow or, equivalently, failing before then. Below we explain how this is done.

Assume that a component's failure can be described by the normal distribution illustrated as a PDF in Figure 9-4. In the graph you'll notice several interesting and useful facts about this component. First, Figure 9-4 shows that 65% of the items will fail at some time within 507 to 693 hours, and 99%

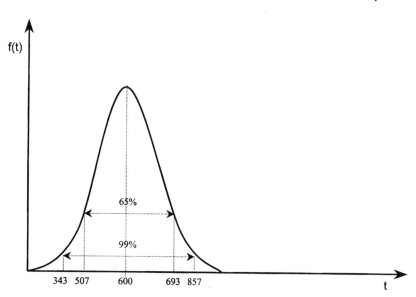

Figure 9-4 The normal (Gaussian) distribution.

will fail between 343 hours and 857 hours. Also, since the PDF is constructed so that the total area under the curve adds up to 1.0 (or 100%), there is a 50% probability that the component will fail before its mean life of 600 hours.

Although some component failure times fit the normal distribution, it is too restrictive for most maintenance situations. For example, the normal distribution is bell-shaped and symmetrical, but often the data is quite skewed. A few items might fail shortly after installation and use, but most survive for a fairly long time. This is depicted in Figure 9-5, which is a Weibull distribution whose shape parameter β equals 2.0. In this case, you can see that the distribution is skewed with a tail to the right. Weibull distribution is popular because it can represent component failures according to the bell-shaped normal distribution, the skewed distribution of Figure 9-5, and many other possibilities. Weibull's equation includes two constants: β (beta), the shape parameter, and η (eta), the char-

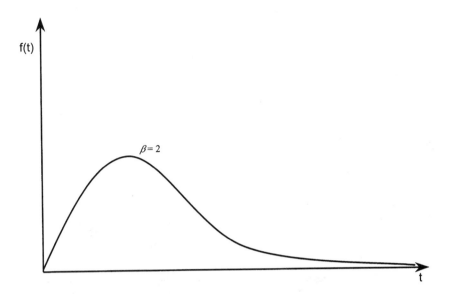

Figure 9-5 Skewed distribution.

acteristic life. It is this design that has made Weibull distribution such a success.

The Hazard Function, depicted in Figure 9-6, clearly shows the risk of a component's failing as it ages. If the failure times have a β value greater than 1.0, the risk increases with age (i.e., it is wearing out). If β is less than 1.0, the risk declines (such as through work hardening or burn-in). If β takes a value equal to 1.0, the failure isn't affected by age (i.e., failures are purely random, caused by external or unusual stress). That is usually the case when a stone hits a car windshield, severely cracking it, which is just as likely to occur with a new car as an old one. In fact, most failures in industrial, manufacturing, process, and transportation industries, when viewed as a system, are random or stress failures.

You want to optimize the maintenance decision, to better know when to replace a component that can fail. If the hazard function is increasing (i.e., β is greater than 1.0), you must

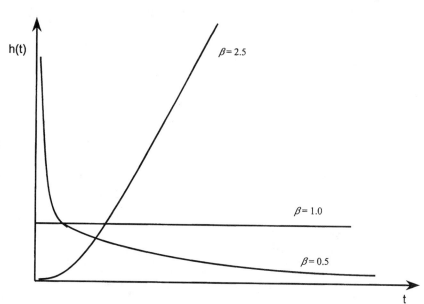

Figure 9-6   Weibull hazard function.

identify where on the increasing risk curve the optimal replacement time occurs. You do this through blending the risk
curve and the costs of preventive maintenance and failure replacement, taking into account component outages for both.
Establishing this optimal time is covered in Chapter 11, "Optimizing Maintenance and Replacement Decisions."

   If the hazard function is constant (β is equal to 1.0) or
declining (β is less than 1.0), your best bet is to let the component run to failure. In other words, preventively replacing
such components will not make the system more reliable. The
only way to do that is through redesign or installing redundancy. Of course, there will be tradeoffs. To make the best
maintenance decision, study the component's failure pattern.
Is it increasing? If so, establish the best time to replace the
component. Is it constant or declining? Then the best action,
assuming there aren't other factors, is to replace the component only when it fails. There isn't any advantage, for either
reliability or cost, in preventive maintenance.

Earlier, we mentioned the importance of reliability software in maintenance management. You can easily establish a component's β value using such standard software. RelCode (3) software was used in Figure 9-7, in which the sample size is 10 with six failure observations and four suspensions, the β value is 3.25, and the mean time to failure of the item is 4760.03 time units. Additional aspects of the figure are covered later in the chapter. Weibull + + (4), M-Analyst (5), and WinSmith (6) perform similar functions.

We must stress that so far we have been focusing on items termed Line Replaceable Units (LRUs). The maintenance action replaces, or renews, the item and returns the component to a statistically as-good-as-new condition. For complex systems with multiple components and failure modes, the form of the hazard function is likely to be the bathtub curve in Figure 9-8. In these cases, the three underlying types of system failure are wearout, quality, and random. Adding them cre-

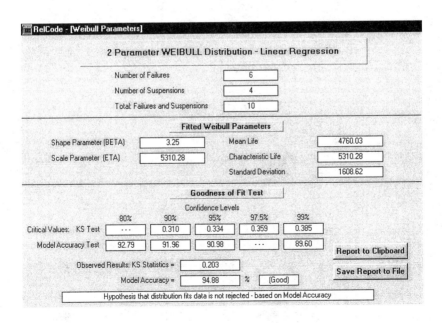

Figure 9-7 RelCode analysis.

Equipment Life Periods

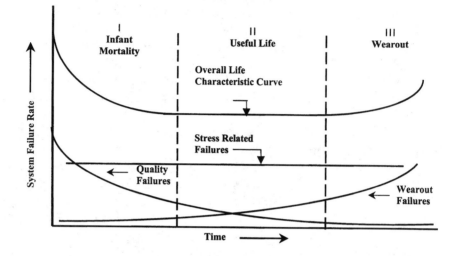

Figure 9-8 System hazard function.

ates the overall bathtub curve. There are three distinct regions: running-in period, regular operation, and wearout, with system failure an increasing risk.

9.5 AN EXAMPLE OF STATISTICAL USE OF FAILURE DATA

Here is an example illustrating how you can extract useful information from failure data. Assume (using the methods to be discussed later in this chapter) that an electrical component has the exponential cumulative distribution function, $F(t) = 1 - e^{-\lambda t}$, where $\lambda = 0.0000004$ failures per hour.

1. What is the probability that one of these parts will fail before 15,000 hours of use?

$$F(t) = 1 - e^{-\lambda t} = 1 - e^{-0.0000004 \times 15000} = 0.006 = 0.6\%$$

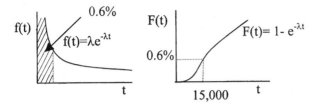

2. How long until you get 1% failures? Rearranging the equation for $F(t)$ to solve for t

$$t = -\frac{\ln[1 - F(t)]}{\lambda} = -\frac{\ln(1 - 0.01)}{0.0000004} = 25,126 \text{ hr}$$

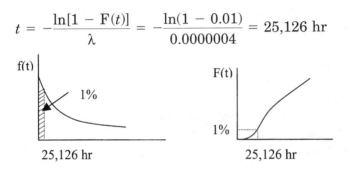

3. What would be the mean time to failure (MTTF)?

$$\text{MTTF} = \int_0^\infty tf(t)dt = \int_0^\infty t\lambda e^{-\lambda t} \, dt = \frac{1}{\lambda} = 2,500,000 \text{ hr}$$

4. What would be the median time to failure (the time when half the number will have failed?

$$F(T_{50}) = 0.5 = 1 - e^{-\lambda T_{50}}$$

$$T_{50} = \frac{\ln 2}{\lambda} = \frac{0.693}{0.0000004} = 1,732,868 \text{ hr}$$

f(t)

50% 50%

Median = 1,732,868 hours

This is the kind of information you can retrieve using the reliability engineering principles in user-friendly software. Read on to discover how.

9.6 REAL-LIFE CONSIDERATIONS: THE DATA PROBLEM

Ironically, reliability management data can slip unnoticed through your fingers as you relentlessly try to control maintenance costs. Data management can prevent that from happening and provide critical information from past experience to improve your current maintenance management process. It is up to upper-level managers, though, to provide trained maintenance professionals with ample computer and technical tools to collect, filter, and process data.

Without doubt, the first step in any forward-looking activity is to get good information. In fact, this is more important than anything else. History proves that progress is built on experience, but there are countless examples where ignoring the past resulted in missed opportunities. Unfortunately, many maintenance departments are guilty of this too.

Companies can be benchmarked against world-class best practices by the extent to which their data effectively guides their maintenance decisions and policies. Here are some examples of how you can make decisions using reliability data management:

1. A maintenance planner notices that an in-service component has failed three times within three months. The superintendent uses this information to estimate the failure numbers in the next quarter, to make sure there will be enough people available to fix them.

2. When ordering spare parts and scheduling maintenance labor, determine how many gearboxes will be

returned to the depot for overhaul for each failure mode in the next year.

3. An effluent-treatment system must be shutdown and overhauled whenever the contaminant level exceeds a toxic limit for more than 60 seconds in a month. Avoid these production interruptions by estimating the level and frequency of preventive maintenance needed.

4. After a design modification to eliminate a problem, determine how many units must be tested, for how long, to verify that the old failure mode has been eliminated, or significantly improved with 90% confidence.

5. A haul truck fleet of transmissions is routinely overhauled at 12,000 hours, as stipulated by the manufacturer. A number of failures occur before the overhaul. Find out how much the overhaul should be advanced or retarded to reduce average operating costs.

6. The cost in lost production is 10 times more than for preventive replacement of a worn component. From this, determine the optimal replacement frequency.

7. You can find valuable information in the database to help with maintenance decisions. For instance, if you know the fluctuating values of iron and lead from quarterly oil analysis of 35 haul truck transmissions, and their failure times over the past three years, you can determine the optimal preventive replacement age (examined in Chapter 12).

Obviously, it is worth your while to obtain and record life data at the system and, where warranted, component levels. When tradespeople replace a component—say, a hydraulic pump—they should indicate which specific pump failed. It may be one of several identical pumps on a complex machine that is critical to operations. They should also specify how it failed, such as "leaking" or "insufficient pressure or volume."

Because we know how many hours equipment operates, we can track the lifetime of individual critical components. That information will then become a part of the company's valuable intellectual asset: the reliability database.

9.7 WEIBULL ANALYSIS

Weibull analysis supported by powerful software is formidable in the hands of a trained analyst. Many examples and comments are given in the practical guidebook *The New Weibull Handbook* (2).

One of the distinct advantages of Weibull analysis is that it can provide accurate failure analysis and forecasts with extremely small samples (2). Let's look closely at the prime statistical failure investigation tool, the Weibull plot. Failure data is plotted on Weibull probability paper but, fortunately, modern software (3–6) provides a "virtual" version. The Weibull plot uses x and y scales, transformed logarithmically so that the Weibull CDF function:

$$F(t) = 1 - e^{-(t/\eta)^\beta}$$

takes the form $y = \beta x +$ Constant. By using Weibull probability "paper," you get a straight line when you plot failure data that fits the Weibull distribution. What's more, the slope of the line will be the Weibull shape parameter β, and the characteristic life η will be the time at which 63.2% of the failures occurred.

9.7.1 Weibull Analysis Steps

Software makes Weibull analysis far more pleasant than it used to be. Conceptually, the software:

- Groups the data in increasing order of time to failure, as in Section 9.2.
- Obtains the median rank from tables for each time

group (the median rank is explained in Section 9.7.3 and the median rank table for up to 12 samples is provided in Appendix 2)

- Plots on Weibull probability paper the median rank versus failure time of each group of observations

For example, in the table in Figure 9-9, the April failures, will have a median rank of 44.83% (Appendix 2), meaning that roughly 44.83% of them occurred up to and including April. The result is a plot such as Figure 9-9. You can see from the graph that the shape parameter β is 1.97 and the characteristic life η is 5.21. Furthermore, the mean life is 4.62 months and the accuracy of how the data fits the Weibull distribution is 99.08%.

Month	Jan	Feb	Mar	Apr	May	June	July	Aug	Sep	Oct
Failures	2	5	7	8	7	6	5	4	3	1
Med rank	3.47	13.84	28.31	44.83	59.30	71.70	22.02	90.30	96.49	98.55

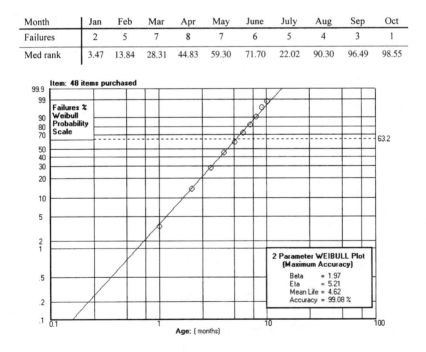

Figure 9-9   Weibull plot.

9.7.2 Advantages

Much can be learned from the plot itself, even how the data deviates from the straight line. For example:

- Whether and how closely the data follows the Weibull distribution
- The type of failure (infant mortality, random, or wearout)
- The component's B_n life (the time at which n% of a population will have failed)
- Whether there may be competing failures (for example, from fatigue and abrasion occurring simultaneously)
- Whether some other distribution, such as log normal, is a better fit
- Whether there may have been predelivery shelf-life degradation
- Whether there is an initial failure-free period that needs to be accounted for
- Forecasts for failures at a given time or during a future given period
- Whether there are batch or lot manufacturing defects

For more information about these deductions, see Ref. 2.

Surprisingly little data, as you will see, is required to draw accurate and useful conclusions. Even with inadequate data, engineers trained to read Weibull plots can learn a lot from them. The horizontal scale is a measure of life or aging such as start/stop cycles, mileage, operating time, takeoff, or mission cycles. The vertical scale is the cumulative percentage failed. The two defining parameters of the Weibull line are the slope β and the characteristic life η.

The characteristic life η, also called the $B_{63.2}$ life, is the age at which 63.2% of the units will have failed. Similarly, where the Weibull line and the 10% CDF horizontal intersects is the age at which 10% of the population fails, or the B_{10} life.

For more serious or catastrophic failures, the B_1, $B_{0.1}$ or $B_{0.01}$ lives are readily obtained from the Weibull plot.

The slope of the Weibull line, β, shows which failure class is occurring. This could be infant mortality (decreasing hazard function, $\beta < 1$), random (constant hazard function, $\beta = 1$), or wearout (increasing hazard function, $\beta > 1$). The $B_{63.2}$ life is approximately equal to the mean-time-to-failure, MTTF. (They are equal when $\beta = 1$. That is when Weibull is equivalent to the exponential distribution.) See Figure 9-10.

A significant advantage of the Weibull plot is that even if you don't immediately get a straight line (such as in Figure 9-9) when plotting the data, you can still learn something quite useful. If the data points are curved downward (Figure 9-11, bottom), it could mean the time origin is not at zero. This could imply that the component had degraded on the shelf, or suffered extended burn-in time after it was made but before delivery. On the other hand, it could show that it was physically impossible for the item to fail early on. Any of these reasons could justify shifting the origin and replotting.

Alternatively, the concave downward curve could be saying that the data really fit a log normal distribution. You can, using software, quickly and conveniently test these hypotheses, replot with a time-origin shift, or transform the scale for log normal probability "paper." Once you apply the appro-

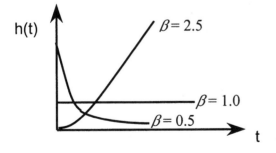

Figure 9-10 Hazard function for burn-in, random, and wearout failures.

priate origin-shift correction (adding or subtracting a value t_0 to each data point time), the resulting plot will be straight and you will know a lot more about the failure process.

Further, if the plotted data forms a dogleg downward (Figure 9-11, top), it could mean that something changed when the failed part was manufactured. The first, steeper-sloped leg reflects low time failures. The second, lower-sloped line indicates failures from batches without the defect. When there are dual failure modes like this, it's known as a batch effect. When there are many suspensions (parts that have been removed for reasons besides failure, or parts that haven't failed at the end of the sampling period) it can be a clue that a batch problem exists. Failure serial numbers clustered closely along a leg support the theory that there was some manufacturing process change before the specific units were produced. Scheduled maintenance can also produce batch effects (2).

An upward-pointing dogleg bend (Figure 9-11, center) indicates multiple failure modes. Investigate the failed parts. If the different failures are plotted separately, treating the other failure as a suspension, two straight lines should be observed. This is the classic bi-Weibull, showing the need for a root cause

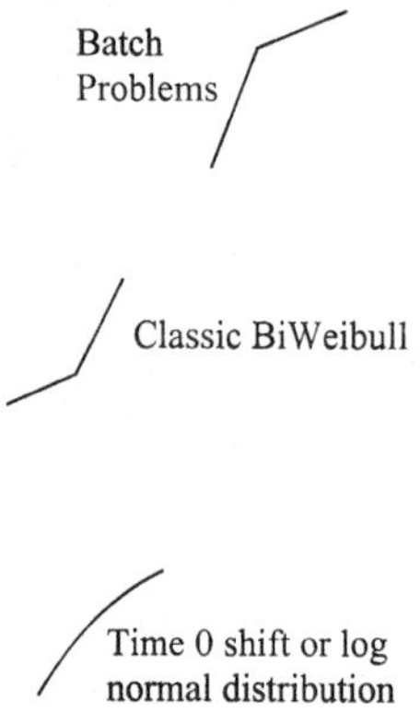

Batch
Problems

Classic BiWeibull

Time 0 shift or log
normal distribution

Figure 9-11   Various Weibull plotted data.

failure analysis (RCFA) to eliminate, for example, an infant-mortality problem. Several dogleg bends distinguish various multiple failure modes (2).

When the Weibull plot forms a downward concave curve, you may need to use an origin shift, t_0, equivalent to using a three-parameter Weibull (2):

$$F(t) = 1 - e^{-(t-t_0/\eta-t_0)^\beta}$$

There should be a physical explanation of why failures cannot occur before t_0. For example, bearings, even in an infant-mortality state, will continue to operate for some time before failing. Use a large sample size, at least 20 failures. If you know, from earlier Weibulls, that the third parameter is appropriate, you might use a smaller sample size, say, eight to ten (2).

For Weibull and other statistical modeling methods, the data requirements are straightforward. Maintenance personnel should be able to collect it during routine activities. There are three criteria for good data, stipulated by D. R. Cox (2):

1. The time origin must be clear.
2. The scale for measuring time must be agreed to.
3. The meaning of failure must be clear.

Although all modern CMMSs (Computerized Maintenance Management Systems) and EAM (Enterprise Asset Management) systems can provide this level of data collection, unfortunately they have been mostly underused. To be fair, writing out failure and repair details isn't part of the tradesperson's job description. Many in the field don't yet see that adding meticulous information to the maintenance system makes organization assets function more reliably. Training in this area could yield untapped benefits.

Although the technicians performing the work provide the most useful information, collecting it is the responsibility of everyone in maintenance. Data must be continually monitored for both quality and relevance. Once the data flow is developed, it can be used in several areas, such as developing preventive maintenance programs, predictive maintenance, warranty administration, tracking vendor performance, and improved decision making (7). As Weibull analysis and other reliability methods become prevalent through advanced software, data will improve. The result? Management and staff will recognize the potential of good data to sharpen their company's competitive edge.

9.7.3 Median Ranks

Suppose that five components fail at 67, 120, 130, 220, and 290 hours. To plot this data on Weibull probability paper, you need the CDF's corresponding estimates. You must estimate the fraction that is failing before each of the failure ages. You can't simply say that the percentage failed at 120 hours is 40 (2/5), because that would imply that the cumulative probability at 290 hours, a random variable, is 100%. This small sample size doesn't justify such a definitive statement. Taken to the absurd, from a sample size of 1 you certainly couldn't conclude that the time of the single failure reflects 100% of total failures. The most popular approach to estimating the Y-axis plotting positions is the median rank. Obtain the CDF plotting values from the median ranks table in Appendix 2, or from a reasonable estimate of the median rank, Benard's formula:

$$\text{Median rank} = \frac{i - 0.3}{n - 0.4}$$

where i = chronological order of the component's failure and n = sample size, 5 in this instance.

Determined by either method, this item's cumulative failure probabilities are 0.13, 0.31, 0.5, 0.69, and 0.87, respectively, for the first, second, third, fourth, and fifth ordered failure observations. When you use reliability software, you do not have to look up the median ranks in tables or perform manual calculations. The program automatically calculates and applies the median ranks to each observation.

9.7.4 Censored Data or Suspensions

It is an unavoidable data-analysis problem that, at the time you are observing and analyzing, not all the units will have failed. You know the age of the unfailed units and that they are still working properly, but not when they will fail. Also, some units may have been replaced preventively. In those cases too, you don't know when they would have failed. These units are said to be suspended or right-censored. Left-censored observations are failures whose starting times are unavailable. While not ideal, statistically you can still use censored data, since you know that the units lasted at least this long. Ignoring censored data would seriously underestimate the item's reliability.

Account for censored data by modifying the order numbers before plotting and determining the CDF values (from Benard's formula or from the median ranks table). Assuming item 2 was removed without having failed, the order numbers 1, 2, 3, 4, and 5 in the previous section (9.7.3) would become: 1, 2.5, 3.75, and 5.625. The formula used to calculate these modified orders is in Appendix 3. Consequently, the modified median ranks would become those shown in Table 9-1. Now

Table 9-1   Modified Orders and Median Ranks for Samples with Suspensions

Hours	Event	Order	Modified order	Median rank
67	F	1	1	0.13
120	S	2	—	—
130	F	3	2.5	0.41
220	F	4	3.75	0.64
290	F	5	5.625	0.99

F, failure; S, suspension (or censored observation).

you can plot the observations on Weibull probability paper normally.

9.7.5 The Three-Parameter Weibull

For the reasons described in Section 9.7.2, including physical considerations, the time of the observations may need to be shifted for the Weibull data to plot to a straight line. By changing the origin, you activate another parameter in the Weibull equation. To make the model work, you must estimate three parameters rather than two. See Appendix 4 for the procedure to estimate the third parameter, also called the location parameter, the guarantee life, or the minimum-life parameter (8). A two-parameter Weibull model is a special case of the more general three-parameter model, with the location parameter equal to zero.

9.7.6 The Five Parameter Bi-Weibull (9)

As shown in Figure 9-11, more than one behavior (of burn-in, random, and wearout) can be reflected by your sample data. The bi-Weibull model can represent two failure types, for example, a stress (random) failure-rate phase followed by a wearout phase; a random phase followed by another random phase, at some higher failure rate; or a burn-in failure followed by random failure. Six common failure patterns are illustrated in Figure 9-12.

A limitation of the Weibull distribution is that it does not cover patterns B and F (Figure 9-12), which often occur (9). A bi-Weibull distribution is formed by combining two Weibull distributions. The five-parameter Weibull, known as the Hastings bi-Weibull, is implemented in the RelCode (3) package. Its hazard function form is:

$$h(t) = \lambda\theta(\lambda\gamma)^{\theta-1} + \left(\frac{\beta}{\eta}\right)\left(\frac{t-\gamma}{\eta}\right)^{\beta-1}$$

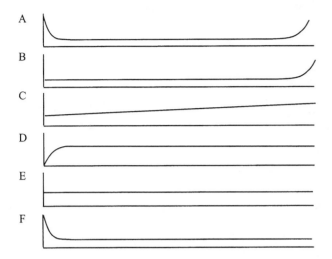

Figure 9-12 System failure rate patterns.

If $\gamma = \theta = 0$, the Hastings bi-Weibull reduces to a Weibull distribution. The bi-Weibull distribution includes the Weibull as a special case, but allows two failure phases, so patterns B and F are now covered. Software (3) will fit the most appropriate of the various models (two-parameter Weibull, three-parameter Weibull, or five-parameter bi-Weibull) to the data. For a random failure phase followed by a wearout phase, knowing the onset of the wearout helps decide when to begin preventive repair or replacement.

9.7.7 Confidence Intervals

You need to be confident that any actions you take based on your statistical analysis and observed data modeling will be successful. In Figure 9-9, we plotted failure observation times of 48 failures over a period of 10 months against the CDF values estimated by the median rank. The median rank estimates that 50% of the time the true percentage of failures lies above or below it.

Similarly, you can estimate the CDF according to another

Table 9-2 Cumulative Probability Distribution Including 95% and 5% Ranks

Month:	Jan	Feb	Mar	Apr	May	Jun	Jul	Aug	Sep	Oct
Failures:	2	5	7	8	7	6	5	4	3	1
Median rank:	3.47	13.84	28.31	44.83	54.30	71.70	82.02	90.30	96.49	98.55
95% rank:	9.51	23.19	39.57	56.60	70.41	81.43	89.85	95.81	99.23	99.89
5% rank:	0.75	7.05	18.57	33.43	47.52	60.43	71.93	81.94	90.49	93.95

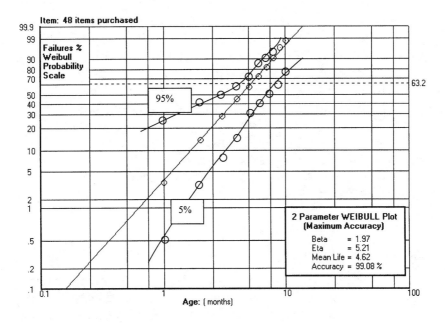

Figure 9-13 Weibull plot showing 5% and 95% confidence limits.

set of tables, the 95% rank tables. This allows that 95% of the time the percentage of failures will be below this value. The same applies for a 5% rank, where 95% of the time the percentage will be above this value. Table 9-2 introduces rows for the 5% and 95% rank values into our original example. The 5% and 95% rank tables are given in Appendix 5. Figure 9-13 shows the data of Table 9-2 on the Weibull plot.

From the Weibull plot of the median, between 95%, and 5% rank lines, you can conclude that the distribution function at the end of May will have a value between 48% and 70%, with a confidence of 90%. This means that, after 5 months, in 90% of the tests you conduct, between 48% and 70% of the items will fail. Or that the reliability (which is 100% minus the CDF value) of this type of item surviving 5 months is between 30% and 52%, 90% of the time.

The vertical distance between the 5% and 95% lines

represents a 90% confidence interval. By plotting all three lines, you get a confidence interval showing that the cumulative probability of failure will assume a value in the interval defined by the intersection of the age and the 5% and 95% lines, with 90% confidence. As a second example, an item's failure probability, up to and including June, is between 60% and 81%, 90% of the time. Or, knowing that $R(t) = 1 - F(t)$, the item's reliability that it will survive to the end of 6 months is between 19% and 40%, with 90% confidence. See Appendix 5 for another example of how this methodology achieves a confidence interval for an item's reliability.

9.7.8 Goodness of Fit

You would, quite naturally, like to have a quantitative measure of how well your model fits the data. How good is the fit? Goodness-of-fit testing provides the answer. Methods such as least-squares or maximum-likelihood (described in Appendix 6) are used to fit an assumed distribution model to the data and estimate the parameters of its distribution function. That information, and the confidence intervals discussed in the previous section, will help you judge the validity of your model choice (be it Weibull, log normal, negative exponential, or another) and your estimation method. One of the methods commonly used is the Kolmogorov-Smirnov test described in Appendix 7.

9.8 ECONOMIC CONSIDERATIONS IN PHYSICAL ASSET MANAGEMENT

Many asset management decisions, such as whether to replace an expensive piece of equipment, require large investments of money. The cost and benefits will continue for a number of years. When an investment today influences future cash flow, keep in mind that the money's value now depends on when it is due. For example, $1000 received in the future is worth less than $1000 received today.

To compare alternative investments, convert the future dollar costs and benefits to their present-day value. In one numerical index, you can summarize the current value of cash flow streams, and rank alternative investments in order of preference (7).

9.8.1 Aspects of Discounting for Capital Equipment Replacement Analysis

If you invest $1000 in a bank account, at an annual interest rate of 10%, after 1 year you will have $1100. If you leave the money in the bank for another year, and the 10% interest rate remains, it will grow to $1210. Discounting uses the same familiar calculation, but in reverse. For example, if you know that in 2 years you will have a maintenance bill of $1210, you can determine the value today of that future cash flow. If the interest rate for discounting is 10%, the current value will be $1000. Details of the calculation are provided in Appendix 8, "Present Value."

The key to discounting is knowing the right interest rate to use in your calculations. Most organizations have clearly stated rates, taking into account the time value of money in capital equipment replacement decisions (7). Rates used include:

- Rate of return on new investments
- Weighted average of market yields on debt and equity
- Cost of additional borrowing
- Rate that keeps the market price of a common stock from falling
- Spread between the prime rate and inflation

The examples in Chapter 11, "Optimizing Maintenance and Replacement Decisions," specify the interest rate to use and explain the calculation method.

Another economic calculation is to convert the present value of a replacement policy to an equivalent annual cost (EAC). This shows the constant value of the cash flow year to year. Some years produce high cash flow—from purchasing new assets, for instance. There are also low-yielding years, such as when maintenance costs are covered by warranty. The EAC smoothes them out into an "average" value year to year for the actual cash flows. See Appendix (9) for a detailed EAC description.

REFERENCES

1. Ernst G. Frankel. Systems Reliability and Risk Analysis. 2nd ed. Norwell, MA: Kluwer Academic, 1983.
2. Robert B. Abernethy. The New Weibull Handbook. 2nd ed. Houston: Gulf Publishing, 1996.
3. Oliver Interactive Inc.: www.oliver-group.com.
4. ReliaSoft: www.weibull.com/home.htm.
5. M-Tech Consulting Engineers: http://www.m-tech.co.za.
6. WinSmith for Windows, Fulton Findings: http://www/weibull news.com.
7. Andrew K.S. Jardine. Optimizing Equipment Maintenance and Replacement Decisions, course notes. Department of Mechanical and Industrial Engineering, University of Toronto.
8. K.C. Kapur, L.R. Lamberson. Reliability in Engineering Design. New York: Wiley, 1977.
9. N.A.J. Hastings. RelCode for Windows User's Manual.

BIBLIOGRAPHY

Fabrycky, W.J., Thuesen, Gerald J. Engineering Economy, 9th ed. Englewood Cliffs, NJ: Prentice Hall, 2001.

Park, C.S., Pelot, R., Porteous, K.C., Zuo, M.J. Contemporary Engineering Economics, 2nd Canadian Edition. Toronto: Addison Wesley Longman, 2000.

RelCode Users' Manual. Albany Interactive, Samford Valley, Queensland, Australia 4520. albany.interactive@bigpond.com

10

Maintenance Optimization Models

Maintenance optimization is all about getting the best result possible, given one or more assumptions. In this chapter, we introduce the concept of optimization through a well understood traveling problem—identifying the best mode of travel, depending on different requirements. We also examine the importance of building mathematical models of maintenance decision problems to help arrive at the best decision.

We look at key maintenance decision areas: component replacement, capital-equipment replacement, inspection procedures, and resource requirements. We use optimization models to find the best possible solution for several problem situations.

Finally, you will learn about the role that artificial intelligence plays in optimization of maintenance decisions.

10.1   WHAT IS OPTIMIZATION ALL ABOUT?

Optimal means the most desirable outcome possible under restricted circumstances. For example, following an RCM study,
you could conduct condition-monitoring maintenance tactics
and/or time-based maintenance and/or time-based discard for
specific parts of a machine or system.

   In this chapter, we introduce maintenance decision optimization. In the following two chapters, we discuss detailed models for asset maintenance and replacement decision making.

   To understand the concept of optimization, consider this
travel routing problem: you have to take an airplane trip, with
three stops, before returning home to Chicago. First destination is London, then Moscow, and finally Hawaii.

   Before purchasing a ticket, you would weigh a number of
options, including airlines, fares, and schedules. You'd make
your decision based on such factors as economy, speed, safety,
and extras.

- If *economy* is most important, you'd choose the airline
  with the cheapest ticket. That would be the optimal
  solution.
- If *speed* is your top priority, you'd consider only the
  schedules and disregard the other criteria.
- If you wanted to optimize *safety*, you'd avoid airlines
  with a dubious safety record and pick only a well-
  regarded carrier.
- If you wanted a *free hotel room* (an "extra") for three
  nights in Hawaii, you'd opt for the airline that would
  provide that benefit.

The above illustrates the concept of optimization. When you
optimize in one area—economy, for example—then you almost always get a less desirable (suboptimal) result in one or
more of the other criteria.

   Sometimes you have to accept a tradeoff between two criteria. For example, although speed may be most important to

you, the cost of traveling on the fastest schedule might be unacceptable. The solution is somewhere in the middle: opting for an acceptable cost (but not the very lowest) and speed (but not the very fastest).

In any optimization situation, including maintenance decision optimization, you should:

- *Think* about optimization when making maintenance decisions.
- Consider *what* maintenance decision you want to optimize.
- Explore *how* you can do this.

10.2 THINKING OPTIMIZATION

Thinking about optimization means considering tradeoffs—the pros and cons. Optimization always has to do with getting the best result where it counts most, while consciously accepting less than that elsewhere.

A Customer Service Manager was asked by the VP of Marketing what he thought his main mission was. His answer: to get every order for every customer delivered without fail on the day the customer specified, 100% of the time. To achieve this goal, the inventory of ready-to-ship goods would have to include every color, size, and style in sufficient quantities to ensure that no matter what was called for, it could be shipped. In spite of unusually big orders, a large number of customers randomly wanting the same thing at the same time, or machinery failure, the manager would have to deliver. His inventory would have had an unacceptably high cost.

The manager failed to realize that a delivery performance just slightly less, say 95%, would be better. In fact, it would be a profit-optimization strategy, the best tradeoff between the cost of inventory and an acceptable and competitive customer-satisfaction level.

10.3   WHAT TO OPTIMIZE

Just as in other areas, you can optimize in maintenance for different criteria, including cost, availability, safety, and profit.

Lowest-costs optimization is often the maintenance goal. The cost of the component or asset, labor, lost production, and perhaps even customer dissatisfaction from delayed deliveries are all considered. Where equipment or component wear-out is a factor, the lowest possible cost is usually achieved by replacing machine parts late enough to get good service out of them but early enough for an acceptable rate of on-the-job failures (to attain a zero rate, you'd probably have to replace parts every day).

Availability can be another optimization goal: getting the right balance between taking equipment out of service for preventive maintenance and suffering outages due to break-downs. If *safety* is most important, you might optimize for the safest possible solution but with an acceptable impact on cost. If you optimize for *profit*, you would take into account not only cost but the effect on revenues through greater customer satisfaction (better profits) or delayed deliveries (lower profits).

10.4   HOW TO OPTIMIZE

One of the main tools in the scientific approach to management decision making is building an evaluative model, usually mathematical, to assess a variety of alternative decisions. Any model is simply a representation of the system under study. When applying quantitative techniques to management problems, we frequently use a symbolic model. The system's relationships are represented by symbols and properties described by mathematical equations.

To understand this model-building approach, examine the following maintenance stores problem. Although simplified, it illustrates two important aspects of model use: con-

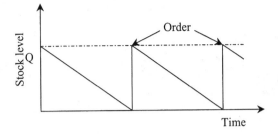

Figure 10-1 An inventory problem.

structing a model of the problem being studied and its solution.

10.4.1 A Stores Problem

A stores controller wants to know how much to order each time the stock level of an item reaches zero. The system is illustrated in Figure 10-1. The conflict here is that the more items ordered at any time, the more ordering costs will decrease but holding costs increase, since more stock is kept on hand. These conflicting features are illustrated in Figure 10-2. The stores controller wants to determine which order quantity will minimize the total cost. This total cost can be plotted, as shown on Figure 10-2, and used to solve the problem.

A much more rapid solution, however, is to construct a mathematical model of the decision situation. The following parameters can be defined:

D = Total annual demand

Q = Order quantity

C_o = Ordering cost per order

C_h = Stockholding cost per item per year

Total cost per year of   = Ordering cost per year +
ordering and holding stock   Stockholding cost per year

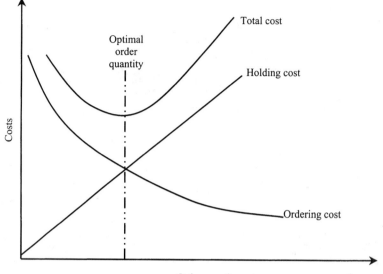

Figure 10-2 Economic order quantity.

Now

Ordering cost per year

= Number of orders placed per year $\times$ Ordering cost per order

$$= \frac{D}{Q} C_o$$

Stockholding cost per year = Average number of items in stock per year (assuming linear decrease of stock) $\times$ Stockholding cost per item

$$= \frac{Q}{2} C_h$$

Therefore, the total cost per year, which is a function of the order quantity and denoted $C(Q)$, is

$$C(Q) = \frac{D}{Q} C_o + \frac{Q}{2} C_h \tag{1}$$

Equation (1) is a mathematical model of the problem relating order quantity Q to total cost C(Q).

The stores controller wants to know the number of items to order to minimize the total cost, i.e., the right-hand side of Equation (1). The answer is obtained by differentiating the equation with respect to Q, the order quantity, and equating the answer to zero as follows:

$$\frac{dC(Q)}{dQ} = -\frac{D}{Q^2} C_o + \frac{C_h}{2} = 0$$

Therefore:

$$\frac{D}{Q^2} C_o = \frac{C_h}{2}$$

$$Q = \sqrt{\frac{2DC_o}{C_h}} \tag{2}$$

Since the values of D, C_o, and C_h are known, substituting them into Equation (2) optimizes the value of the order quantity Q. Strictly check that the value of Q obtained from Equation 2 is a minimum and not a maximum. You can ensure that this is the case by taking the second derivative of C(Q) and noting the positive result. In fact, the order quantity equalizes the holding and ordering costs per year.

Example: Let D = 1000 items, C_o = \$5, C_h = \$0.25:

$$Q = \sqrt{\frac{2 \times 1000 \times 5}{0.25}} = 200 \text{ items}$$

Each time the stock level reaches zero, the stores controller should order 200 items to minimize the total cost per year of holding and ordering stock.

Note that various assumptions have been made in the inventory model presented, which may not be realistic. For example, no consideration has been given to quantity discounts, the possible lead time between placing an order and its receipt, and the fact that demand may not be linear or known for certain. The purpose of the above model is simply to illustrate constructing a model and attaining a solution for a particular problem. There is abundant literature about stock-control problems without many of these limitations. If you are interested in stock-control aspects of maintenance stores, see Nahmias (1).

It's clear from the above inventory control example that we need the right kind of data, properly organized. Most organizations have a Computerized Maintenance Management System (CMMS) or an Enterprise Asset Management (EAM) system. The vast amount of data they store makes optimization analyses possible.

Instead of building mathematical models of maintenance decision problems, software is available to help you make optimal maintenance decisions. This is covered in Chapters 11 and 12.

10.5 KEY MAINTENANCE MANAGEMENT DECISION AREAS

There are four key decision areas that maintenance managers must address to optimize their organization's human and physical resources. These areas are depicted in Figure 10-3. The first column deals with component replacement, the second with establishing the economic life of capital equipment, the third with inspection decisions, including condition monitoring. The final column addresses decisions concerning resources required for maintenance and their location. To build strong maintenance optimization, you need an appropriate source, or sources, of data. The foundations for this, as shown in Figure 10-3, are the CMMS and the EAM system.

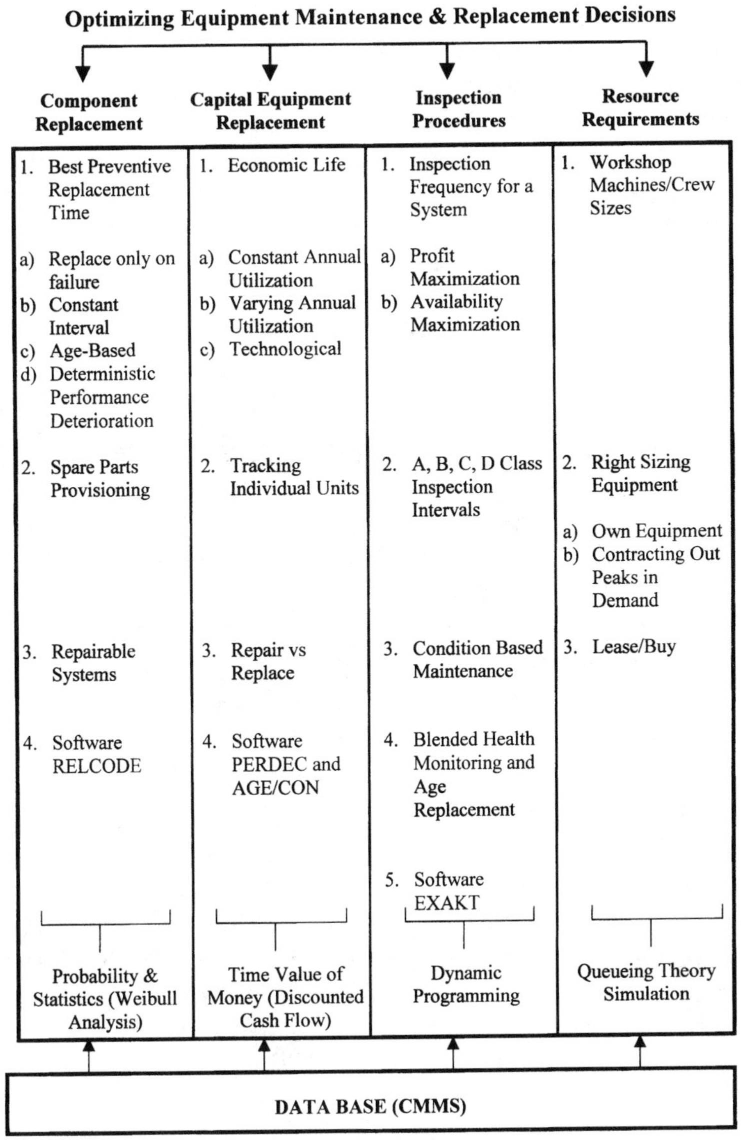

Optimizing Equipment Maintenance & Replacement Decisions

Component Replacement	Capital Equipment Replacement	Inspection Procedures	Resource Requirements
1. Best Preventive Replacement Time	1. Economic Life	1. Inspection Frequency for a System	1. Workshop Machines/Crew Sizes
a) Replace only on failure b) Constant Interval c) Age-Based d) Deterministic Performance Deterioration	a) Constant Annual Utilization b) Varying Annual Utilization c) Technological	a) Profit Maximization b) Availability Maximization	
2. Spare Parts Provisioning	2. Tracking Individual Units	2. A, B, C, D Class Inspection Intervals	2. Right Sizing Equipment
			a) Own Equipment b) Contracting Out Peaks in Demand
3. Repairable Systems	3. Repair vs Replace	3. Condition Based Maintenance	3. Lease/Buy
4. Software RELCODE	4. Software PERDEC and AGE/CON	4. Blended Health Monitoring and Age Replacement	
		5. Software EXAKT	
Probability & Statistics (Weibull Analysis)	Time Value of Money (Discounted Cash Flow)	Dynamic Programming	Queueing Theory Simulation

DATA BASE (CMMS)

Figure 10-3 Maintenance decision optimization.

In Chapters 11 and 12 we discuss optimization of key maintenance decisions in component replacement, capital equipment replacement, and inspection procedures (columns 1, 2, and 3 of the framework illustrated in Figure 10-3). The framework's foundation, or database, is addressed in detail in Chapter 4. In this chapter, we focus on column 4, resource requirements, and comment briefly on the role of artificial intelligence in maintenance optimization.

10.5.1 Resource Requirements

When it comes to maintenance resource requirements, you must decide what resources there should be, where they should be located, who should own them, and how they should be used.

If sufficient resources aren't available, your maintenance customers will be dissatisfied. Having too many resources, though, isn't economical. Your challenge is to balance spending on maintenance resources such as equipment, spares, and staff with an appropriate return for that investment.

10.5.2 Role of Queuing Theory to Establish Resource Requirements

The branch of mathematics known as queuing theory, or waiting-line theory, is valuable in situations where bottlenecks can occur. You can explore the consequences of alternative resource levels to identify the best option. Figure 10-4 illustrates the benefit of using queuing theory to establish the optimal number of lathes for a workshop. In this example the objective is to ensure that the total cost of owning and operating the lathes and tying up jobs in the workshop is minimized. For a model of this decision process, see Appendix 10.

10.5.3 Optimizing Maintenance Schedules

In deciding maintenance resource requirements, you must also consider how to use resources efficiently. An important

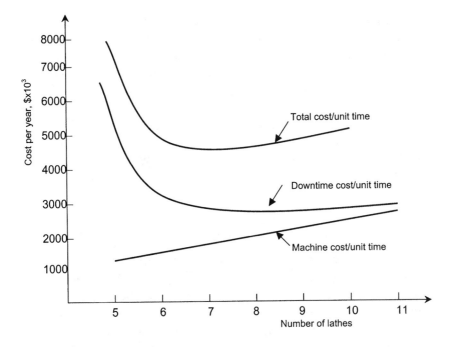

Figure 10-4 Optimal number of machines in a workshop.

consideration is scheduling jobs through a workshop. If there is restricted workshop capacity and jobs cannot be contracted out, you must decide which job should be done first when a workshop machine becomes available.

Sriskandarajah et al. (2) present a unique and highly challenging maintenance overhaul scheduling problem at the Hong Kong Mass Transit Railway Corporation (MTRC). In this case, preventive maintenance keeps trains in a specified condition, taking into account both the maintenance cost and equipment-failure consequences. Since maintenance tasks done either "too early" or "too late" can be costly, how maintenance activities were scheduled was important.

Smart scheduling reduces the overall maintenance budget. In establishing a schedule, it's essential to acknowledge constraints, such as the number of equipment/machines that can be maintained simultaneously, as well as economic, reliability, and technological concerns. Most scheduling problems are a combination of factors. Because of this complexity, a heuristic technique known as genetic algorithms (GAs) was used in this case to arrive at the global optimum.

The performance of the algorithm compared well with manual schedules established by the Hong Kong MTRC. For example, total costs were reduced by about 25%. The study supports the view that genetic algorithms can provide good solutions to difficult maintenance scheduling problems.

10.5.4 Optimal Use of Contractors
(Alternative Service Delivery Providers)

The above decision process assumed that all the maintenance work had to be done within the railway company's own maintenance facility. If you can contract out the maintenance tasks, you must decide, for instance, whether to contract out all, some, only during peak demand periods, or none of them. The conflicts of making such decisions are illustrated in Figure 10-5.

If your organization doesn't have any maintenance resources, you will need to contract out all the work. The only cost will be paying for the alternative delivery service. If your organization has a maintenance division, though, there will be two cost components. One is a fixed cost for the facility, shown in Figure 10-5 to increase linearly as the facility size expands. The second cost is variable, increasing as more work is handled internally but leveling off when there is overcapacity.

The optimal decision is a balance between internal resources and contracting out. Of course, this isn't always the best solution. You have to assess the demand pattern, and both internal and external maintenance costs. If it is cheaper not to have an internal maintenance facility, the optimal solution would be to contract out all the maintenance work. Alternatively, the optimal solution could be to gear up your own

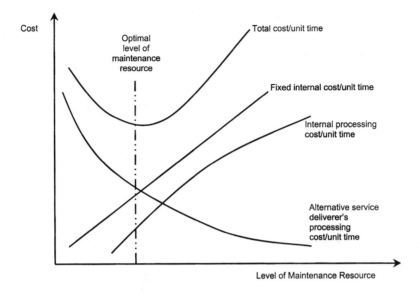

Figure 10-5   Optimal contracting-out decision.

organization to do all the work internally. See Appendix 11 for a model of this decision process.

10.5.5   Role of Simulation in Maintenance Optimization

Common maintenance management concerns about resources include:

- How large should the maintenance crews be?
- What mix of machines should there be in a workshop?
- What rules should be used to schedule work through the workshop?
- What skill sets should we have in the maintenance teams?

Some of these questions can be answered by using a mathematical model. For complex cases, though, a simulation model of the decision situation is often built. You can use the simulation to evaluate a variety of alternatives, then choose the best. Many simulation software packages are available that require minimal programming.

Many resource decisions are complex. Take, for example, a situation where equipment in a petrochemical plant requires attention. If the maintenance crew is limited by size or attending to other tasks, the new job may be delayed while another emergency crew is called in. Of course, if there had been a large crew in place initially, this wouldn't have occurred. Since there can be many competing demands on the maintenance crew, it's difficult to establish the best crew size.

This is where simulation is valuable. Figure 10-6 illustrates a case in which simulation was used to establish the

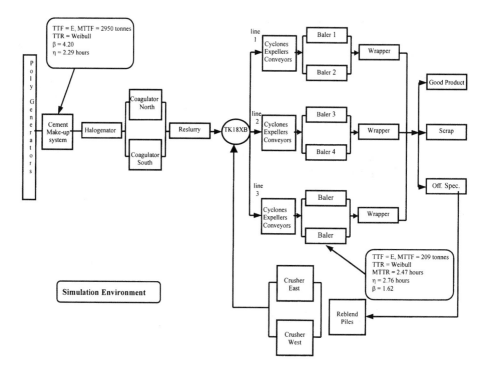

Figure 10-6 Maintenance crew size simulation.

optimal maintenance crew size and shift pattern in a petrochemical plant. For illustration purposes, statistics are provided for the failure pattern of select machines, along with repair-time information. The study was undertaken to increase the plant's output. The initial plan was to add a fourth finishing line, to reduce the bottleneck at the large mixing tank (TK18XB), just before the finishing lines. The throughput data, though, showed that once a machine failed, there was often a long wait until the maintenance crew arrived. Rather than construct another finishing line, it was decided to increase maintenance resources. By using simulation, the throughput from the plant was significantly increased with little additional cost. Increasing crew size was a much cheaper alternative and achieved the required increased throughput.

10.6 ROLE OF ARTIFICIAL INTELLIGENCE IN MAINTENANCE OPTIMIZATION: EXPERT SYSTEMS AND NEURAL NETWORKS

Collecting data for maintenance decision making is the relatively easy part. There are many hardware and software products with impressive accuracy, capacity, and computer interfaces on the market. The result is oceans of data. However, analyzing the data is more difficult, because human experts are required to interpret it. Two technological solutions are in varying evolutionary states: expert systems and neural networks.

10.6.1 Expert Systems

Table 10-1 shows the kind of experience and engineering know-how you will find in expert system knowledge bases. It presents the likelihood that a particular problem is related to the pattern of vibration irregularities by frequency order.

Table 10-1 An Example of an Expert System Knowledge Base

Fault	Probabilities (%)					
	1x rpm	2x rpm	>2x	≫2x	Other	<1x
Shaft imbalance	90	5	5			
Shaft bent	90	5	5			
Lost rotor vanes	90	5	5			
Case distortion	60	20	10			10
Foundation distortion	40	30			10	20
Bearing damage	40	20				20
Gear damage			20	60	20	
Bearing casing loose	30				10	90
Bearing liner loose					10	90
Bearing eccentricity	40	60				
Casing structural resonance	70	10				10
Coupling damage	20	30	10		80	20
Rotor resonance	100					
Rotor looseness	10	10	10		10	40
Shaft misalignment	30	60	10			

Expert systems are a branch of artificial intelligence. They can be used to process and interpret large numbers of data, including trends, ratios, and other combinations of signals. Figure 10-7 depicts a rule that is part of an expert system knowledge base on oil analysis.

Expert systems consist of rules, possibly thousands, such as those in Figure 10-7, that "fire" in sequence to arrive at a set of final recommendations. When all the conditions in the IF part of the rule are true, the rule fires. Each statement, called a choice, in the THEN and ELSE part of a rule is weighted (3). The same choices can appear in many rules. Their weights are totaled and divided by the number of times they appear in rules that have fired. A choice will appear in the set of final recommendations if its average weight exceeds some predefined limit, say 8/10. A fuzzy expert system is an expert system that uses a collection of fuzzy membership functions and rules, instead of Boolean logic, to reason about data (described in Appendix 12).

> **Rule:**
> **IF Machine type is (engine diesel)**
> **AND condition 1 (Iron high)**
> **AND condition 2 (Chromium high)**
> **AND ...**
> **THEN**
> **Choice 1 diagnostic (PISTON RING WEAR/DAMAGE) – 7/10**
> **Choice 2 recommendation (Replace Piston Ring within 100 hours) – 9/10**
> **AND**
> **...**
> **ELSE**
> **Choice 3, etc**

Figure 10-7   An oil-analysis expert system rule.

10.6.2 Neural Networks

A disadvantage of expert systems is their dependence on human knowledge. Programming the expert system with human experience can be time-consuming and error-prone. Another approach is to use neural networks, the other main branch of artificial intelligence. Unlike expert systems, they do not require that human knowledge be programmed into them, because they observe and learn on their own during a training period.

Each of the nodes in Figure 10-8 is called a neuron. This type of neural network is known as a multilayer perceptron (MLP). The neurons of an MLP are arranged in layers. The bottom layer is the input layer. The top one is the output layer. There can be any number of middle or hidden layers. The lines connecting the nodes are known as synaptic links, whose values or weightings change as the neural net is trained.

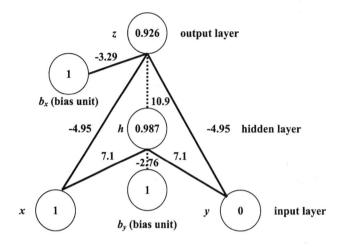

Figure 10-8 A neural network.

Figure 10-8 represents a three-layer neural network, providing XOR (the Boolean "exclusive or") functionality. After a suitable training period the values of the synaptic links are as shown. To get an answer, first compute the sum of inputs to the hidden layer neuron h: $s = (1 \times 7.1) + (1 \times -2.76) + (0 \times 7.1) = 4.34$. Then apply an activation function $v = 1/(1 + e^{-s})$ to compute the activation value, $v = 1/(1 + e^{-4.34}) = 0.98\%$. The output neuron is computed similarly: $s = (1 \times -4.95) + (0 \times -4.95) + (0.98\% \times 10.9) + (1 \times -3.29) = 2.52$ and $v = 0.926$. Of course, 0.91 is not quite 1 but for this example, is close enough. Appendix 13 shows how this particular neural net is trained. You will find more practical information on neural networks in Ref. 4.

A Condition Maintenance Fault Diagnosis (CMFD) system can use an MLP classifier to determine the state of a plant from a measurement pattern. The classifier chooses a class from among a number of predefined classes (e.g., Normal, Fault A, Fault B, etc). Each potential condition is represented by an output neuron.

Training neural networks is like "curve fitting." Think of

it as a form of nonlinear regression (4). One limitation is that you may need an astronomical number of training cycles for problems with many inputs, such as those related to condition-based maintenance. Nonetheless, active research is reporting new successful applications (5). Among them, neural nets have become popular in cameras and cars, in which the automatic transmission learns your driving style during the first few hundred miles.

REFERENCES

1. S. Nahmias. Production and Operations Analysis. 3rd ed. Irwin/McGraw-Hill, 1997.
2. C. Sriskandarajah, A. K. S. Jardine, C. K. Chan. Maintenance Scheduling of Rolling Stock Using a Genetic Algorithm. Journal of the Operational Research Society 1998; 49:1130–1145.
3. www.exsys.com.
4. Donald R. Tveter. Manual to the pro version of Backprop Algorithm software. http://www.dontveter.com/bpr/bptutinps.zip and http://www.dontveter.com/bpr/public2.html.
5. M. H. Jones and D. G. Sleeman, eds. Condition Monitoring '99. Coxmoor Publishing, 1999.

11

Optimizing Maintenance and Replacement Decisions

In this chapter, we explore the various strategies and tools you need to make the best maintenance and replacement decisions. In particular, you need to know the optimal replacement time for critical system components (also known as line-replaceable units, or LRUs), and capital equipment.

At the LRU level, we examine age- and block replacement strategies. You'll learn how RelCode software can help you optimize LRU maintenance decisions to keep costs under control and increase equipment availability.

With capital equipment, it's critical to establish economic viability. We examine how to do this in two operating environments:

- Constant use, year by year
- Declining use, year by year (older equipment is used less)

In this chapter, you'll discover how to extend the life of capital equipment through a major repair or rebuild, and when that is more economical than replacing it with new equipment. The optimal decision is the one that minimizes the long-run Equivalent Annual Cost (EAC), the life-cycle cost of owning, using, and disposing of the asset. Our study of capital-equipment replacement includes a description of AGE/CON and PERDEC software, which simplify the job of managing assets.

Finally, we address the importance of inspection frequency, including a case study for a fleet of buses subject to A, B, C, and D inspection levels.

11.1   INTRODUCTION: ENHANCING RELIABILITY THROUGH PREVENTIVE REPLACEMENT

Generally, preventing a maintenance problem is always preferred to fixing it after the fact. You'll increase your equipment reliability by learning to replace critical components at the optimal time, before a breakdown occurs. When is the best time? That depends on your overall objective. Do you most want to minimize costs or maximize availability? Sometimes the best preventive replacement time accomplishes both objectives, but not necessarily.

Before you start, you need to obtain and analyze data to identify the best preventive replacement time. Later in this section, we present some maintenance policy models of fixed interval and age-based replacements to help you do that.

For preventive replacement to work, these two conditions must first be present:

1.   If cost is most important, the total cost of a replacement must be greater after failure than before. If reducing total downtime is most essential, the total downtime of a failure replacement must be greater

than a preventive replacement. In practice, this usually happens.

2. The risk of a component failing must increase as it ages. How can you know? Check that the Weibull shape parameter associated with the component's failure times is greater than 1. It is often assumed that this condition is met, but be sure that is in fact the case. See Chapter 9 for a detailed description of Weibull analysis.

11.1.1 Block Replacement Policy

The block replacement policy is sometimes called the group or constant interval policy, since preventive replacement occurs at fixed times and failure replacements whenever necessary. The policy is illustrated in Figure 11-1. C_p is the total cost of a preventive replacement, C_f is the total cost of a failure replacement, and t_p is the interval between preventive replacements.

You can see that for the first cycle of t_p, there isn't a failure, while there are two in the second cycle and none in the third and fourth. As the interval between preventive replacements decreases, there will be fewer failures in between. You want to obtain the best balance between the investment in preventive replacements and the economic consequences of failure replacements. This conflict is illustrated in Figure 11-2. $C(t_p)$ is the total cost per week of preventive replacements occurring at intervals of length t_p, with failure replace-

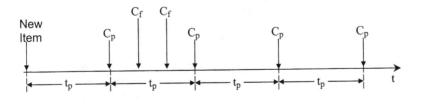

Figure 11-1 Constant-interval replacement policy.

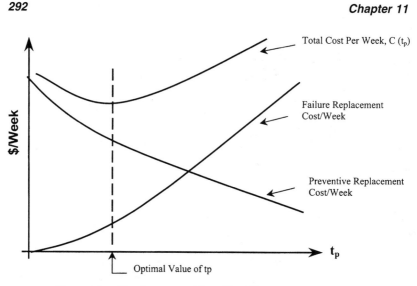

Preventive Replacement Cost Conflicts

Figure 11-2   Constant-interval policy: optimal replacement time.

ments occurring whenever necessary. See Appendix 14 for the total cost curve equation.

The following problem is solved using RelCode software (1), incorporating both the cost model and Weibull analysis to establish the best preventive replacement interval.

11.1.1.1   Problem

A bearing has failed in the blower used in diesel engines. The failure has been established according to a Weibull distribution, with a mean life of 10,000 km and a standard deviation of 4500 km. The bearing failure is expensive, costing 10 times as much to replace than if it had been done as a preventive measure. Determine:

- The optimal preventive replacement interval (or block policy) to minimize total cost per kilometer
- The expected cost saving of the optimal policy over a run-to-failure replacement policy

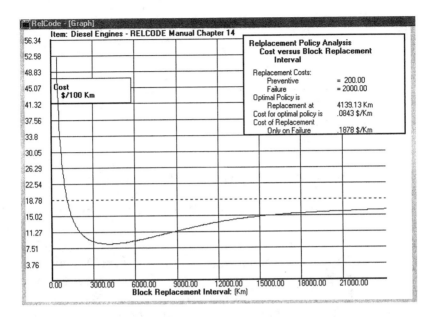

Figure 11-3 RelCode output: block replacement.

- Given that the cost of a failure is $2000, the cost per km of the optimal policy

11.1.1.2 Result

Figure 11-3 shows a screen capture from RelCode. You can see that the optimal preventive replacement time is 4140 kilometers (4139.13 in the chart). The figure also provides valuable additional information. For example:

- The cost per kilometer of the best policy is $0.0843/km
- The cost saving compared to a run-to-failure policy is $0.1035/km (55.11%)

11.1.2 Age-Based Replacement Policy

In the age-based policy, the preventive replacement time depends on the component's age—if a failure replacement occurs, the component's time clock is reset to zero. In the block

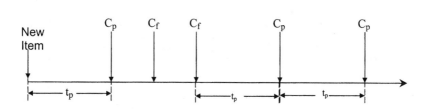

Figure 11-4 Age-based replacement policy.

replacement policy, however, preventive replacements occur at fixed intervals regardless of the operating component's age. In this case, the component is replaced only when it reaches the specified age.

Figure 11-4 illustrates the age-based policy. C_p and C_f represent the block replacement policy, and t_p represents the component age when preventive replacement occurs. You can see that there aren't any failures in the first cycle, which is t_p. After the first preventive replacement, the component must have failed before reaching its next planned preventive replacement age. After the first failure replacement, the clock is set to zero and the next preventive replacement scheduled. However, before it's reached, the component is again replaced due to failure. After this second failure replacement, the component survives to the planned preventive replacement age of t_p. Similarly, the next replacement cycle shows that the component made it to its planned preventive age.

The conflicting cost consequences are identical to those in Figure 11-2, except that the x-axis measures the actual age (or utilization) of the item, rather than a fixed time interval. See Appendix 15 for the mathematical model depicting this preventive replacement policy.

The following problem is solved using RelCode software, which incorporates the cost model, to establish the best preventive replacement age.

11.1.2.1 Problem

A sugar-refinery centrifuge is a complex machine composed of many parts that can fail suddenly. It's decided that the plow-setting blade needs preventive replacement. Based on the age-based policy, replacements are needed when the setting blade reaches a specified age. Otherwise, a costlier replacement will be needed when the part fails. Consider the optimal policy to minimize the total cost per hour associated with preventive and failure replacements.

To solve the problem, you have the following data:

- The labor and material cost of a preventive or failure replacement is $2000.
- The value of production losses is $1000 for a preventive replacement and $7000 for a failure replacement.
- The failure distribution of the setting blade can be described adequately by a Weibull distribution with a mean life of 152 hours and a standard deviation of 30 hours.

11.1.2.2 Result

Figure 11-5 shows a screen capture from RelCode. As you can see, the optimal preventive replacement age is 112 hours (111.77 hours in the figure), and there's additional key information that you can use. For example, the preventive replacement policy costs 45.13% of a run-to-failure policy [(59.19 − 32.48)/59.1] × 100, making the benefits very clear. Also, the total cost curve is fairly flat in the 90 to 125 hours region, providing a flexible planning schedule for preventive replacements.

11.1.3 When to Use Block Replacement

On the face of it, age replacement seems to be the only sensible choice. Why replace a recently installed component that is still working properly? In age replacement, the component always remains in service until its scheduled preventive replacement age.

To implement an age-based replacement policy, though, you must keep an ongoing record of the component's current age and change the planned replacement time if it fails. Clearly, the cost of this is justified for expensive components,

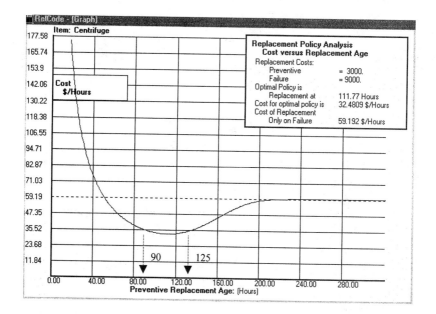

Figure 11-5 RelCode output: age-based replacement.

but for an inexpensive one the easily implemented block replacement policy is often more cost-efficient.

11.1.4 Safety Constraints

With both block and age replacement, the objective is to establish the best time to preventively replace a component, to minimize the total cost of preventive and failure replacements. If you want to ensure that the failure probability doesn't exceed a particular value—say, 5%—without cost considerations being formally taken into account, you can determine when to schedule a preventive replacement from the cumulative failure distribution. This is illustrated in Figure 11-6, in which you can see that the preventive replacement should be planned once the item is at 4000 km.

Alternatively, you can preventively replace a critical component so that the risk, or hazard, doesn't exceed a specified

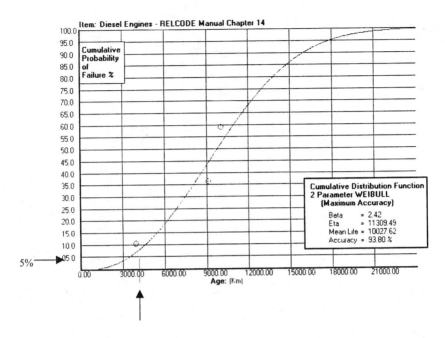

Figure 11-6 Optimal preventive replacement age: risk-based maintenance.

value, such as 5×10^{-5} failures per km. In this case, you need to use the hazard plot to identify the preventive replacement age. This is illustrated in Figure 11-7, which shows that the component's appropriate preventive replacement age is 4500 km.

11.1.5 Minimizing Cost and Maximizing Availability

In both block and age replacement, the objective is minimizing cost. Maximizing availability simply requires replacing the total costs of preventive and failure replacement in the models with their total downtime. Minimizing total downtime is then equivalent to maximizing availability. In Appendix 16, you'll find total downtime-minimization models for both block and age replacement.

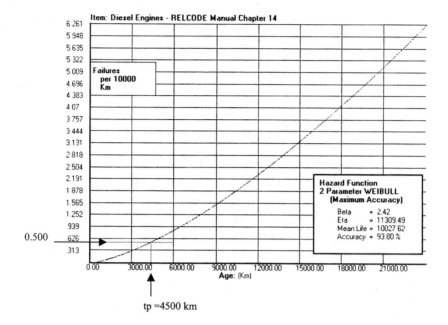

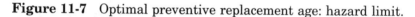

Figure 11-7 Optimal preventive replacement age: hazard limit.

11.2 DEALING WITH REPAIRABLE SYSTEMS

Rather than completely replace a failed unit, you may be able to get it operating with minor corrective action. This is what's known as a minimal repair. It's assumed that the unit's overall hazard rate hasn't been changed, that the maintenance fix has simply made it operational again, and that the previous hazard level has remained the same. This is illustrated in Figure 11-8, where $h(t)$ is the component's hazard rate at age t.

On the other hand, you could decide to take significant maintenance action when an item fails. This is called a general repair in maintenance and is illustrated in Figure 11-9. You can see that a general repair reduces the unit's hazard rate, but not to zero, which would be the case if you completely renewed it. Taking into account the costs of minimal repair (a

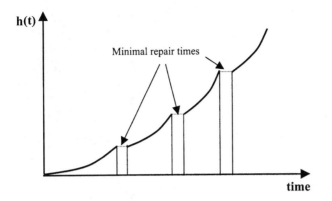

Figure 11-8   Minimal repair.

minor fix), general repair (such as an overhaul), and a complete renewal, you want to know the best maintenance action when a unit fails.

This is an area that is not well developed, although progress is being made. One approach is illustrated in Figure 11-10. Virtual age, just after the last repair, is measured on the x-axis and length of operating time (run time) since last repair is measured on the y-axis, at the top of the figure. The

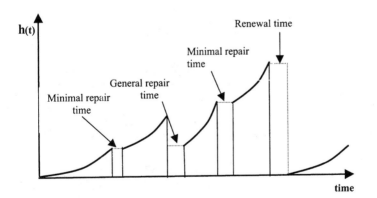

Figure 11-9   Minimal and general repair.

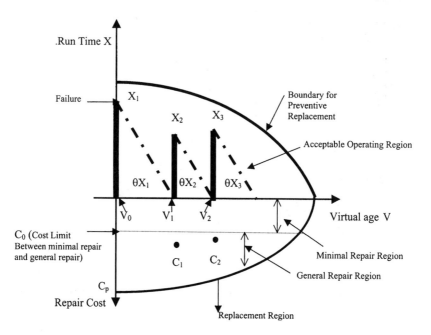

Figure 11-10   Repairable system optimization.

optimal policy for this system is explained through the different regions shown in the figure.

When a failure occurs, if the repair cost—shown on the y-axis in the bottom of the figure—is less than a constant value C_0, the maintenance is minimal repair. Otherwise, it is a general repair, provided the cost doesn't exceed the lower boundary. If it does exceed the lower boundary, you must replace the system. You should preventively replace the system whenever the length of operating time (run time) reaches the upper boundary. Otherwise, the system is in the safe operating region (2).

In Figure 11-10, you can see that at the start the unit was of virtual age V_0. At first failure time (X_1), the cost was C_1, resulting in a general repair and the equipment's virtual age being V_1. At failure time X_2, the repair cost was C_2 and, again, there is a general repair resulting in the unit's virtual age becoming V_2.

11.3 ENHANCING RELIABILITY THROUGH ASSET REPLACEMENT

Eventually, it becomes economically justifiable to replace an aging asset with a new one. Since it's usually years between replacements, rather than weeks or months as it often is for component preventive replacement, you must consider the fact that money changes in value over time. This is known as discounted cash-flow analysis. Figure 11-11 shows the different cash flows when replacing an asset on a 1-, 2-, and 3-year cycle.

To decide which of the three alternatives would be best, you must compare them fairly. You can do this by converting all cash flows associated with each cycle to today's prices, or their current value. In Chapter 9, "Reliability Management and Maintenance Optimization: Basic Statistics and Economics," we examined the process of discounting cash flows. Also,

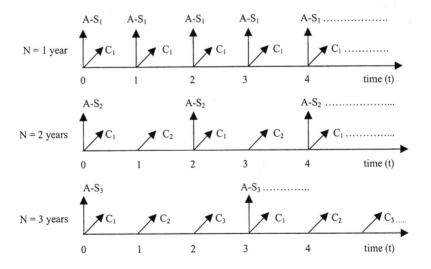

Figure 11-11 Asset-replacement cycles.

to be fair, you must compare all possible cycles for the alternatives over the same period of time, often infinity. This may seem unrealistic, but it isn't; it actually keeps the analysis straightforward. The process is used in the following section.

11.3.1 Economic Life of Capital Equipment

There are two key conflicts in establishing the economic life of capital equipment:

- The increasing operations and maintenance costs of the aging asset
- The declining ownership cost in keeping the asset in service, since the initial capital cost is being written off over a longer time period

These conflicts are illustrated in Figure 11-12, in which fixed costs (such as operator and insurance charges) are also depicted by the horizontal line.

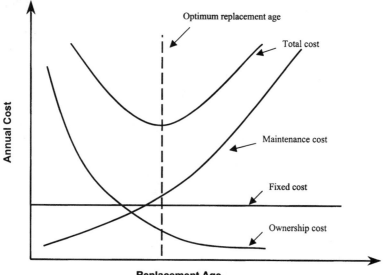

Figure 11-12 Economic-life problem.

The asset's economic life is the time at which the total cost is minimized. You will find the total cost curve equation, along with its derivation, in Appendix 17. Rather than rely directly on the economic-life model, you can use software that incorporates it. The following problem is solved using PERDEC software (3), which contains the economic-life model provided in Appendix 18.

11.3.1.1 Problem

Canmade Inc. wants to determine the optimal replacement age for its materials handling turret side loaders, to minimize total discounted cost. Historical data analysis has produced the information in Table 11-1 (all costs are in present-day prices). The cost of a new turret side loader is $150,000 and the interest rate, for discounting purposes, is 12%.

11.3.1.2 Solution

Using PERDEC produces Figure 11-13, which shows that the economic life of a side loader is 3 years, with an annual cost of $79,973. This amount would be sufficient to buy, operate, and sell side loaders on a 3-year cycle. This is the optimal decision. Note that the amount in the annual budget to fund replacements would be calculated based on the number and age of side loaders and based on a 3-year replacement cycle.

Although the above example has considered only O&M costs and resale value, you must be sure to include all relevant

Table 11-1 Sample Historical Data for Determining Life Cycle of Equipment

Year	Average operating and maintenance cost ($/year)	Resale value at end of year ($)
1	16,000	100,000
2	28,000	60,000
3	46,000	50,000
4	70,000	20,000

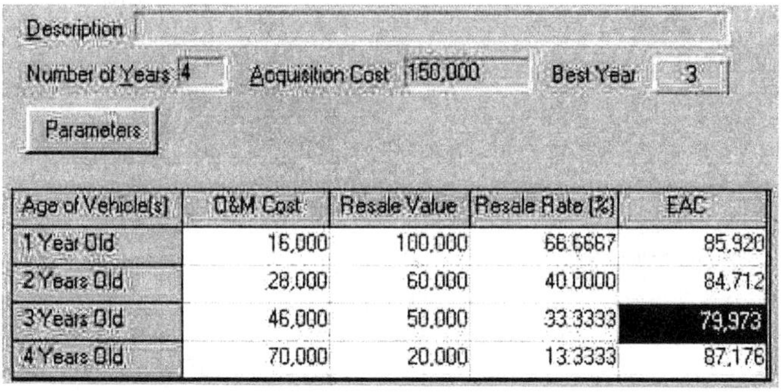

Description				
Number of Years 4	Acquisition Cost 150,000		Best Year	3
Parameters				

Age of Vehicle[s]	O&M Cost	Resale Value	Resale Rate [%]	EAC
1 Year Old	16,000	100,000	66.6667	85,920
2 Years Old	28,000	60,000	40.0000	84,712
3 Years Old	46,000	50,000	33.3333	79,973
4 Years Old	70,000	20,000	13.3333	87,176

Figure 11-13 PERDEC optimal replacement age.

costs. Figure 11-14, taken from an American Public Works Association publication (3), shows that, for example, the cost of holding spares (inventory) and downtime is included in the analysis.

The calculation using the "classic" economic-life model assumes that equipment is being replaced with similar equipment. It also assumes that the equipment is used steadily year by year. You may find this useful for some analyses, such as forklift-truck replacement, if the equipment's design isn't impacted significantly and its use is constant.

Some equipment isn't used steadily year to year. You might use new equipment frequently and older equipment only to meet peak demands. In this case, you need to modify the classic economic model and examine the total cost of the group of similar equipment rather than individual units. Examples where this applies include:

- Machine tools, when new tools are heavily used to meet basic workload and older tools used to meet peak demands, say, during annual plant shutdowns.
- Materials handling equipment, such as older forklift trucks in a bottling plant, when they are kept to meet seasonal peak demands.

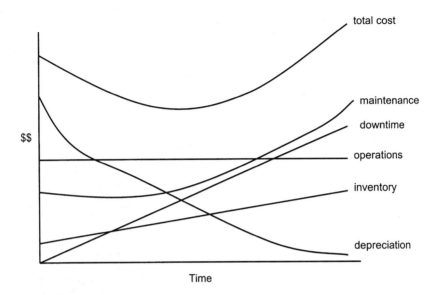

Figure 11-14 APWA economic-life model: establishing the economic life of equipment whose utilization varies during its life. (From Ref. 4.)

- Trucking fleets undertaking both long-distance and local deliveries. New trucks are used on long haul routes initially, then, as they age, are relegated to local deliveries.

The following is an example of establishing the economic life of a small fleet of delivery vehicles, using AGE/CON software (3).

11.3.1.3 Problem

A company has a fleet of eight vehicles to deliver their products to customers. The company uses its newest vehicles during normal demand periods and the older ones to meet peak demands. In total, the fleet travels 100,000 miles per year, and these miles are distributed, on average, among the eight vehicles as follows:

Vehicle 1 travels 23,300 miles/year
Vehicle 2 travels 19,234 miles/year
Vehicle 3 travels 15,876 miles/year
Vehicle 4 travels 15,134 miles/year
Vehicle 5 travels 12,689 miles/year
Vehicle 6 travels 8,756 miles/year
Vehicle 7 travels 3,422 miles/year
Vehicle 8 travels 1,589 miles/year

Determine the optimal replacement age for this class of delivery vehicle.

11.3.1.4   Solution

You must first establish how often a vehicle is used as it ages. (For the underlying mathematical model of when best to replace aging equipment, see Appendix 19. It features a case study that establishes the economic replacement policy for a large fleet of urban buses.) The utilization data will look like Figure 11-15. The trend in this figure can be described by the equation of a straight line:

$$Y = a - bX$$

where Y is miles/year and X is the vehicle number—vehicle 1 is the most utilized; 8 is the least utilized. Using the actual figures given above, you can establish from AGE/CON (or by plotting the data on graph paper or using a trend-fitting software package) that the equation in this case would read:

$$Y = 26,152 - 3034X$$

Next you must establish the trend for Operating and Maintenance (O&M) costs. For vehicle 1, (your newest vehicle, and the one used the most) you need the following information:

Miles traveled last year (already given)   23,300
O&M cost—say, for last year $3150

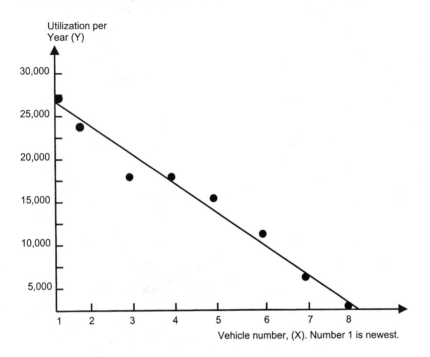

Figure 11-15   Vehicle-utilization trend.

<table>
<tr><td>Cumulative miles on the odometer to
the midpoint of last year</td><td>32,000</td></tr>
</table>

As you can see, the O&M cost/mile is $0.14 for vehicle 1. Do the same for all eight vehicles. Vehicle 8 (the oldest, and the one least used) may have data that looks like this:

<table>
<tr><td>Miles traveled last year (already given)</td><td>1589</td></tr>
<tr><td>O&M—say, for last year</td><td>$765</td></tr>
<tr><td>Cumulative miles on the odometer to
the midpoint of last year</td><td>120,000</td></tr>
</table>

In this case, the O&M cost/mile is $0.48 for vehicle 8. A plot of the trend in O&M cost would look like Figure 11-16. Each vehicle's O&M cost is represented by a dot on the graph. Vehicles 1 and 8 are identified in the diagram. The straight line

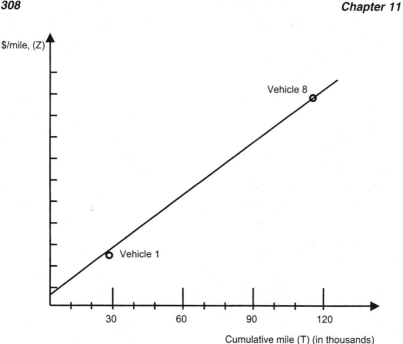

Figure 11-16   Trend in O&M cost.

is the trend that has been fitted to the dots. The equation you
get this time is:

$$Z = 0.0164 + 0.00000394T$$

where Z = \$/mile and T = cumulative miles traveled. The two
trend lines are both used as input to AGE/CON. Note that in
both cases, a straight (linear) relationship existed for $Y(X)$ and
$Z(T)$ so that the fitted lines read $Y = a - bX$ and $Z = c + dT$.
Often a polynomial equation will give a better fit to a particular series of data. These polynomial equations can be generated by using a standard statistical package such as Minitab
or SPSS.

   To solve the problem, you require additional information.
Assume that a new delivery vehicle costs \$40,000. The resale
values for this particular type of vehicle are:

1-year-old vehicle: $28,000
2-year-old vehicle: $20,000
3-year-old vehicle: $13,000
4-year-old vehicle: $6,000

The interest rate for discounting purposes is 13%.

Figure 11-17, from AGE/CON, shows that the optimal replacement age of a delivery vehicle is 4 years with an associated EAC of $14,235. To implement this recommendation, you'd likely need to replace a quarter of the fleet each year, so the same number of vehicles would be replaced each year. All vehicles, then, would be replaced at the end of their fourth year of life.

11.3.2 Before- and After-Tax Calculations

In most cases, you conduct economic-life calculations on a before-tax basis. This is always the case in the public sector, where tax considerations are not applicable. In the private sector, your financial group can help decide the best course of

AGE/CON - Main

File   Edit   View   Parameters   Help

Description Delivery Vehicle

Number of Years 4   Acquisition Cost 40,000   Best Year 4

Parameters

Age of Vehicle(s)	O&M Costs	Resale Value	Resale Rate [%]	EAC
1 Year Old	1,432	28,000	70.0000	18,818
2 Years Old	2,405	20,000	50.0000	16,724
3 Years Old	2,156	13,000	32.5000	15,348
4 Years Old	1,118	6,000	15.0000	14,236

Figure 11-17 AGE/CON optimal replacement age.

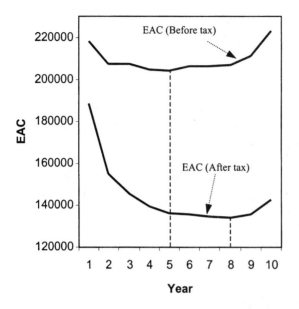

Figure 11-18 Economic life: before and after tax calculation (Feller Buncher data).

action. In many cases, the after-tax calculation doesn't alter the decision, although the EAC is reduced when the result is in after-tax dollars.

Figure 11-18 illustrates what happened when Feller Buncher was replaced in the forestry industry. The data used is provided in Tables 11-2 and 11-3. You can see from the figure that the general trend of the EAC curve remains the same, but it veers downward when the tax implications are included in the economic-life model. In before-tax dollars the economic

Table 11-2   Feller Buncher: Base Data

Acquisition cost of Feller Buncher	$526,000
Discount rate	10%
Capital cost allowance (CCA)	30%
Corporation tax (CT)	40%

Table 11-3   Feller Buncher: Annual Data

Year	O&M cost	Resale value
1	8332	368,200
2	60,097	275,139
3	107,259	212,423
4	116,189	169,939
5	113,958	104,189
6	182,516	95,085
7	173,631	85,981
8	183,883	83,958
9	224,899	73,842
10	330,375	40,462

life of the Feller Buncher is 5 years, while in after-tax dollars the minimum is 8 years. Note that in both cases the total cost curve is flat around the minimum. In this example, a replacement age between 5 and 8 years would be good for either a before-tax or an after-tax calculation.

In Appendix 20, you'll find the form of the EAC model for after-tax, taking into account corporation tax and capital-cost allowance. The model is in AGE/CON, which provided the graphs in Figure 11-18.

If you are making after-tax calculations, take care that all relevant taxes and current rules are incorporated into the model. One example is the report by Buttimore and Lim (5) that deals with the cost of replacing shovels in the mining industry on an after-tax basis. It includes not only corporation tax and capital-cost allowances, but federal and provincial taxes applicable to the mining industry at the time.

11.3.3   The Repair-Versus-Replacement Decision

You may be facing a sudden major maintenance expenditure for equipment, perhaps due to an accident. Or you may be able to extend the life of an asset through a major overhaul. In either case, you have to decide whether to make the maintenance expenditure or dispose of the asset and replace it with a new one. The example below illustrates an approach that can help you make the most cost-effective decision.

11.3.3.1 Problem

A major piece of mobile equipment, a front-end loader used in open-pit mining, is 8 years old. It can remain operational for another 3 years, with a rebuild costing $390,000. The alternative is to purchase a new unit costing $1,083,233. Which is the better alternative?

11.3.3.2 Solution

You need additional data to make an informed decision that will minimize the long-run EAC. Review historical maintenance records for the equipment to help forecast what the future O&M costs will be after rebuild, and obtain O&M cost estimates and tradein values from the supplier of the potential new purchase.

Figure 11-19 depicts the cash flows from acquiring new equipment at time T where:

- R is the cost of the rebuild
- $C_{p,i}$ is the estimated O&M cost of using the present equipment after rebuild in year i; $i = 1, 2, \ldots$ T
- A is the cost of acquiring and installing new equipment

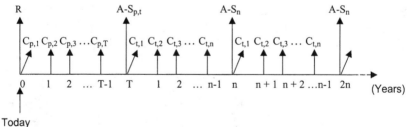

Figure 11-19 Cash flows associated with acquiring new equipment at time T.

Table 11-4   Cost Data: Current
Equipment

$C_{p,1} = \$138,592$	$S_{p,0} = \$300,000$
$C_{p,2} = \$238,033$	$S_{p,1} = \$400,000$
$C_{p,3} = \$282,033$	$S_{p,2} = \$350,000$
	$S_{p,3} = \$325,000$

- T is the time at which the changeover occurs from the present equipment to new equipment; T = 0, 1, 2, 3
- $S_{p,T}$ is the tradein value of the present equipment at the changeover time T
- $C_{t,j}$ is the estimated O&M cost associated with using the new equipment in year j; j = 1, 2, . . . n
- S_n is the tradein value of the new equipment at age n years
- n is the economic replacement age of the new equipment

The necessary data for the current equipment is provided in Table 11-4 and for the new equipment in Table 11-5. The interest rate for discounting is 11% and the purchase price for the new equipment is: A = \$1,083,233. Evaluating the data gives the new equipment an economic life of 11 years, with an associated EAC of \$494,073.

11.3.3.3   Solution

To decide whether to rebuild, calculate the EAC associated with not rebuilding (i.e., T = 0), replacing immediately, re-

Table 11-5   Cost Data: New Equipment

$C_{t,1} = \$38,188$	$S_1 = \$742,500$
$C_{t,2} = \$218,583$	$S_2 = \$624,000$
$C_{t,3} = \$443,593$	$S_3 = \$588,000$
$C_{t,4} = \$238,830$	$S_4 = \$450,000$
Etc.	Etc.

Table 11-6   Optimal Changeover Time

	Changeover time to new loader, T			
	T = 0	T = 1	T = 2	T = 3
Overall EAC ($)	449,074	456,744	444,334	435,237 (minimum)

building then replacing after 1 year (i.e., T = 1), and rebuilding then replacing after 2 years (i.e., T = 2), rebuilding and
replacing after 3 years (i.e., T = 3), and so on. The result of
these various options is shown in Table 11-6. The solution is
to rebuild and then plan to replace the equipment in 3 years,
at a minimum EAC of $435,237.

Note: In a full study, the rebuilt equipment might be kept
for a longer time. See Appendix 21 for the model used to conduct the above analysis.

11.3.4   Technological Improvement

If a new, more technically advanced model of equipment you
are using becomes available, you will have to weigh the costs
and benefits of upgrading. See Appendix 22 for a basic model
to evaluate whether or not to switch. The case study by Buttimore and Lim (5) dealing with shovel replacement in an open-
pit mining operation is an extension of the model in Appendix
22, with a technologically improved design also having a better productivity rate.

11.3.5   Life-Cycle Costing

Life-cycle costing (LCC) analysis considers all costs associated
with an asset's life cycle, which may include:

- Research and development
- Manufacturing and installation
- Operation and maintenance
- System retirement and phaseout

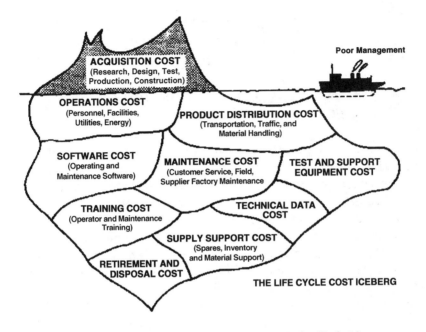

Figure 11-20 The life-cycle cost iceberg. (From Ref. 6.)

Essentially, when making decisions about capital equipment—be it for replacement or new acquisition—reflect on all associated costs. Figure 11-20 (6) is a good illustration of what can be involved. The iceberg shows very well that, while costs like up-front acquisition are obvious, the total cost can be many times greater. In the airline industry, the life-cycle cost of an aircraft can be five times its initial purchase. An advertisement by Compaq Computer Corp. states that 85% of computer costs are usually hidden, going to administration (14%), operations (15%), and the bulk—about 56%—to asset management and service and support.

In the economic-life examples covered in this chapter, we take an LCC approach, including costs for:

- Purchase
- O&M
- Disposal value

Of course, if necessary, you may have to include other costs that contribute to the LCC, such as spare-parts inventory and software maintenance.

11.4 ENHANCING RELIABILITY THROUGH INSPECTION

A big part of the maintenance mandate, of course, is to ensure that the system is reliable. One way to achieve this is to identify critical components that are likely to fail and preventively replace them. In Section 11.1, we covered methods to identify which components (line-replaceable units) are candidates for preventive replacement, and how to decide when to do it.

An alternative is to consider the system as a whole and make regular inspections to identify problem situations. Then, carry out minor maintenance, such as changing a component or topping up the gearbox with oil to prevent system failure. You need to know, though, the best frequency of inspection (Section 11.4.1).

Yet another approach is to monitor the health of the system through predictive maintenance and act only when you get a signal that a defect is about to happen that, if not corrected, will cause a system failure. This third approach is covered in detail in the next chapter.

11.4.1 Establishing the Optimal Inspection Frequency

Figure 11-21 shows a system composed of five components, each having its own failure distribution. (In Reliability-Centered Maintenance terminology, these components are different modes of system failure.)

Because the Weibull failure pattern is so flexible, all the components can likely be described as failing according to a Weibull distribution, but with different shape parameters. That is, depending on the component, the risk of it failing as it ages can either increase, remain constant, or decrease. The

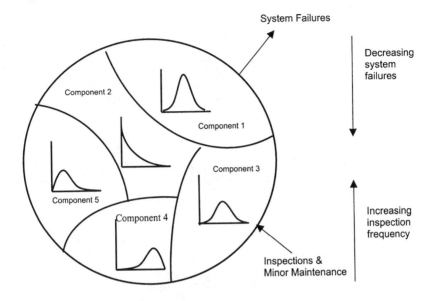

Figure 11-21   System failures.

overall effect is that there will be system failures from any number of causes. If you analyze the overall system failure data, again a Weibull would fit. The shape parameter almost certainly would equal 1.0, though, indicating that system failures are occurring strictly randomly. This is what you should expect. The superposition of numerous failure distributions creates an exponential failure pattern, the same as a Weibull with shape parameter of 1.0 (7).

To reduce these system failures, you can inject inspections, with minor maintenance work, into the system, as shown in Figure 11-21. The question is then: How frequently should inspections occur? While you may not know the individual risk curves of system components, you can determine the overall system failure rate by examining the maintenance records. The pattern will almost certainly look like Figure 11-22, in which the system failure rate is constant but can be reduced through increasing the inspection frequency. The risk curve

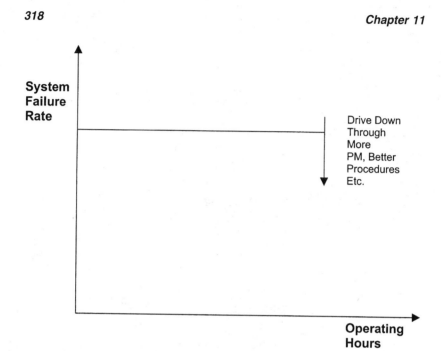

Figure 11-22   System failure rate.

may not be constant if system failures emanate from one main cause. In this case, the system-failure distribution will be identical to the component's failure pattern.

Another way of viewing the situation is to consider the system's Mean Time to Failure (MTTF). As the system failures decline, the MTTF will increase. This is illustrated in the probability density functions in Figure 11-23.

If the optimal inspection frequency minimizes the total downtime of the system, you get the conflicting curves of Figure 11-24. See Appendix 23 for the underlying mathematical model that establishes the optimal frequency.

11.4.1.1   Case Study

An urban transit fleet had a policy of inspecting the buses every 5000 km. At each inspection, an A, B, C, or D check was made. An A check is a minor maintenance inspection; a D

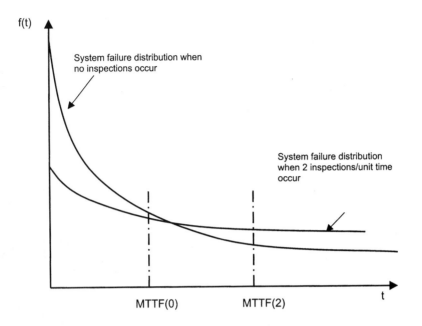

Figure 11-23 Inspection frequency versus MTTF.

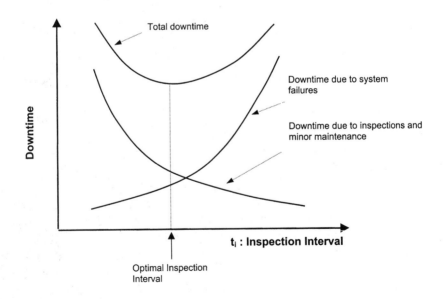

Figure 11-24 Optimal inspection frequency.

Table 11-7   Transit Commission's Bus Inspection Policy

km (1000)	Inspection type			
	A	B	C	D
5	×			
10		×		
15	×			
20			×	
25	×			
30		×		
35	×			
40			×	
45	×			
50		×		
55	×			
60			×	
65	×			
70		×		
75	×			
80				×
Total	8	4	3	1	$\Sigma = 16$
R_i	0.5	0.25	0.1875	0.0625	$\Sigma = 1.0$

R_i = no. of type i inspections/total no. of inspections.
i = A, B, C, or D.

check is the most detailed. The policy is illustrated in Table 11-7.

Since the 5000-km inspection policy was not followed precisely, some inspections took place before 5000 km, and others were delayed. As a result, as shown in Figure 11-25, the average distance traveled by a bus between failures decreased as the interval between the checks was increased.

Knowing the average time a bus was out of service due to repair and inspection established the optimal inspection interval at 8000 km, as Figure 11-26 shows. Note that the total downtime curve is very flat around the optimum, so the current policy of making inspections at the easily implemented 5000-km interval might be best. Before the analysis, though,

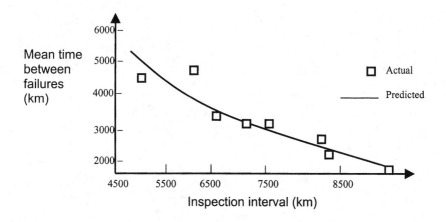

Figure 11-25   Mean distance to failure versus inspection interval.

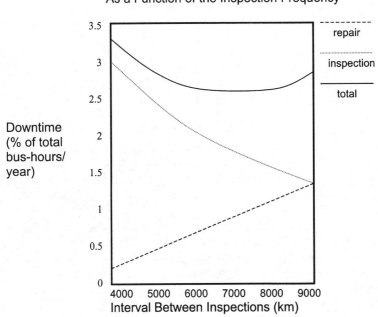

Figure 11-26   Optimal inspection interval.

it wasn't known whether the current policy was appropriate or should be modified. As you can see, data-driven analysis revealed the answer. The study by Jardine and Hassounah (8) provides full details.

11.5 SOFTWARE THAT OPTIMIZES MAINTENANCE AND REPLACEMENT DECISIONS

In this chapter, we have referred to several software packages: RelCode for component-replacement decisions, AGE/CON for mobile-equipment replacement, and PERDEC for fixed-equipment replacement. There are other good software tools on the market, and more becoming available all the time, all of which you can find on the Web.

REFERENCES

1. RelCode, Albany Interactive, Stamford Valley, Queensland, Australia. albany.interactive@bigpond.com
2. X. Jiang, V. Makis, A. K. S. Jardine. Optimal repair/replacement policy for a general repair model. Advances in Applied Probability, 2001; Vol. 33.
3. www.oliver-group.com
4. H. Green, R. E. Knorr. Managing Public Equipment. The APWA Research Foundation, 1989:105–129.
5. B. Buttimore, A. Lim. Noranda equipment replacement system. In: Applied Systems and Cybernetics. Vol 2. Pergamon Press, 1981:1069–1073.
6. B. S. Blanchard. Logistics Engineering and Management. 5th ed. Englewood Cliffs, NJ: Prentice Hall, 1998.
7. R. F. Drenick. The failure law of complex equipment. Journal of the Society of Industrial Applied Math 1960; 8:680–690.
8. A. K. S. Jardine, M. I. Hassounah. An optimal vehicle-fleet inspection schedule. Journal of the Operations Research Society 1990; 41(9):791–799.

12

Optimizing Condition-Based Maintenance

The best time for preventive repairs or machinery and component replacements is just before a failure occurs. That's only logical, but it can be hard to judge. In this chapter, we explore condition-based maintenance (CBM), an effective new approach to help you extend the usefulness of your equipment, and overcome the challenge of predicting when it will fail.

We discuss the difficulty of assessing the large volume of data (known as condition data) captured by such proactive maintenance tactics as oil and vibration analysis and the benefits of Proportional Hazards Modeling. This statistical technique connects condition data to the corresponding component failure age, using EXAKT™ software (1). Its use is described in seven steps starting with preparing, or *cleaning*, the data the model will be based on.

After building the PHM, we take you through the process of conducting a statistical confidence test to determine how well the model fits the data. Once that's done, you calculate the transition probabilities between each state (equipment condition). Then the various hypotheses are tested to fine-tune the model. That's followed by comparing the preventive repair costs with those when a system or component fails. The resulting decision model defines the optimal maintenance policy that applies each time you receive and process new condition data. The decision model includes an estimate of the equipment's remaining useful life (RUL) and recommends whether maintenance should be done immediately.

Finally, we show how a sensitivity analysis is performed. The objective is to determine how much the optimal replacement policy will be affected by operational changes in maintenance costs.

12.1 INTRODUCTION

CBM is obviously a good idea. Of course, you want the maximum useful life from each physical asset before taking it out of service for preventive maintenance. But translating this into an effective monitoring program, and timely maintenance, isn't necessarily easy. It can be difficult to select the monitoring parameters most likely to indicate the machine's state of health, and then to interpret the influence of those measured values on the machinery's RUL. We address these problems in this chapter.

The essential questions to pose when implementing a CBM program are:

- Why monitor?
- What equipment components need to be monitored?
- What monitoring technologies should be used?
- How (according to what signals) should the monitoring be done?

- When (how often) should the monitoring be done?
- How should the condition monitoring results be interpreted and acted on?

Reliability-Centered Maintenance (RCM), described in Chapter 7, helps to find the right answers to the first three questions. However, additional optimizing methods are required to handle the remainder. In Chapters 9 and 10, you learned that the way to approach these problems is to build a model describing the factors surrounding maintaining or replacing equipment. In Chapter 10, we dealt with the lifetime of components considered independent random variables, meaning that no additional information, other than equipment age, is used to schedule preventive maintenance.

CBM, however, introduces new information, called condition data, to determine more precisely the most advantageous moment at which to make a repair or replacement. We extend the models in Chapter 10 to include the influence of condition data on the RUL of machinery and its components. The extended modeling method we introduce in this chapter, which takes measured data into account, is known as Proportional Hazards Modeling (PHM), and the measured condition data are referred to as covariates.

Since D. R. Cox's pioneering paper on PHM was published in 1972 (cited in Ref. 2), it has been used primarily to analyze medical survival data. Since 1985, it has been used more extensively, including application to marine gas turbines, nuclear reactors, aircraft engines, and disk brakes on high-speed trains. In 1995, A. K. S. Jardine and V. Makis at the University of Toronto initiated the CBM Laboratory (3) to develop general-purpose software applying proportional hazards models to available maintenance data. The software was designed to be integrated into a plant's maintenance information system to optimize its CBM activities. The result, in 1997, was a program called EXAKT™, now in its second version and rapidly earning attention as a CBM-optimizing methodology. We produced the examples in this chapter, with their graphs and calculations, using the EXAKT program. To work through the examples using the program, contact the author for a demonstration version (4).

Industry has adopted various monitoring methods that produce a signal when a failure is about to occur. The most common are vibration monitoring and oil analysis. Moubray (5) provides an overview of condition-monitoring techniques, including the following:

- Vibration analysis
- Ultrasonic analysis
- Ferrography
- Magnetic chip detection
- Atomic emission spectroscopy
- Infrared spectroscopy
- Gas chromatography
- Moisture monitoring
- Liquid dye penetrants
- Magnetic particle inspection
- Power signature analysis

As Pottinger and Sutton (6) said, "much condition-monitoring information tells us something is not quite right, but it does not necessarily inform us what margins remain for exploitation before real risk, and particularly real commercial risk, occurs." In this chapter, you will learn how to accurately estimate equipment health using condition monitoring. The goal is to make the optimal maintenance decision, blending economic considerations with estimated risk.

Early work on estimating equipment risk dealt with jet engines on aircraft (7). Oil analysis was conducted weekly and, if unacceptable metal levels were found in the samples, the engine was removed before its scheduled removal time of 15,000 flying hours. A PHM was constructed by statistically analyzing the condition data, along with the corresponding age of the engines that functioned for the duration and those that were removed due to failure (in this case, operating outside tolerance specifications). Three key factors, of a possible 20, emerged for estimating the risk that the engine would fail:

- Age of engine
- Parts per million of iron (Fe) in the oil sample at the time of inspection
- Parts per million of chrome (Cr) in the oil sample at the time of inspection

The PHM also identifies the weighting for each risk factor. The complete equation used to estimate risk of the jet engine failing was:

$$\text{Risk at time of inspection} = \frac{4.47}{24,100}\left(\frac{t}{24,100}\right)^{3.47} e^{0.41\text{Fe}+0.98\text{Cr}}$$

where the contribution of the engine age toward the overall failure risk is:

$$\frac{4.47}{24,100}\left(\frac{t}{24,100}\right)^{3.47}$$

(this is termed a Weibull baseline failure rate), and the contribution to overall risk from the risk factors from the oil analysis is $e^{0.41\text{Fe}+0.98\text{Cr}}$.

The constants in the age-contribution portion of the risk model 4.47, 24,100, and 3.47 are obtained from the data and will change depending on the equipment. They may even be different for the same equipment if operating in a different environment. The key iron and chrome risk factors are also equipment- and operating environment–specific. In this case, it was the absolute values of iron and chrome that were used. In other cases, rates of change or cumulative values may be more meaningful for risk estimation. By carefully analyzing condition-monitoring data, along with information about the age and reason for equipment replacement, you can construct an excellent risk model.

Optimizing maintenance decisions usually requires that more than just risk of failure be taken into account. You may want to maximize the operating profit and/or equipment availability or minimize total cost. In this chapter, we assume your objective is to minimize the total long-term cost of preventive and failure maintenance. Besides determining the risk curve, then, you must get cost estimates for both prevention and failure replacement, and failure consequences.

Being able to detect failure modes, which gradually lessen functional performance, can dramatically impact overall costs. This, therefore, is the first level of defense in the RCM task planning logic as illustrated in Figure 12-1. The

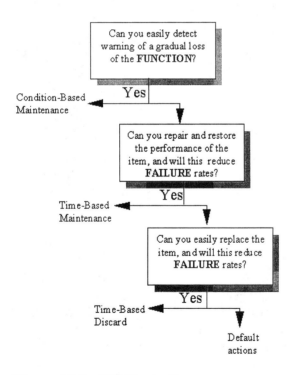

Figure 12-1 RCM logic diagram.

diagram shows that CBM is preferred if the impending failure can be "easily" detected in ample time. This proactive intervention is illustrated in the P-F interval (8) in Figure 12-2. Investigating the relationship between past condition surveillance and past failure data helps develop future maintenance management policy and specific maintenance decisions (9). In addition, modern, flexible maintenance information-management systems compile and report performance, cost, repair, and condition data in numerous ways.

Maintenance engineers, planners, and managers perform CBM by collecting data that reflects the state of equipment or component health. These condition indicators (or covariates) can take various forms. They may be continuous, such as operational temperature or feed rate of raw materials. They may

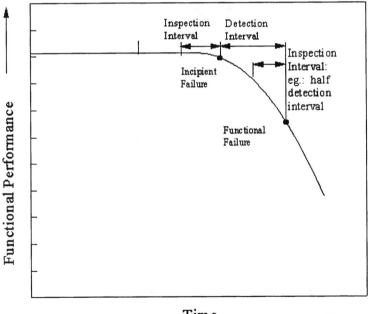

Figure 12-2 The P-F interval.

be discrete, such as vibration- or oil-analysis measurements (2). They may be mathematical combinations or transformations of the measured data, such as measurement change rates, rolling averages, and ratios. Since it's hard to know exactly why failures occur, the condition indicator choices are endless. Without a systematic means of judging and rejecting superfluous data, you will find CBM far less useful than it should be. PHM is an effective approach to dealing with information overload because, based on substantial historic condition and corresponding failure-age data, it can decipher the equipment's current condition and make an optimal recommendation.

In this chapter, we describe, with examples, each step of the PHM process. An integral part of the process is statistical testing of various hypotheses. This helps avoid the trap of blindly following a method without adequately verifying whether the model suits the situation and data.

We divide the problem of optimizing CBM into seven steps:

1. Preparing and studying the data
2. Estimating the parameters of the Proportional Hazards Model
3. Testing how "good" the PHM is
4. Defining states (designating, for example, low, medium, and high ranges) and determining the transition probabilities from one state to another
5. Testing various assumptions concerning transition behavior in different time intervals and the interdependence of covariate groups
6. Making the optimal decision for lowest, long-run maintenance cost
7. Analyzing the sensitivity of the optimal decision policy to operational cost changes

These steps are graphically represented in the software's interactive flow diagram in Figure 12-3.

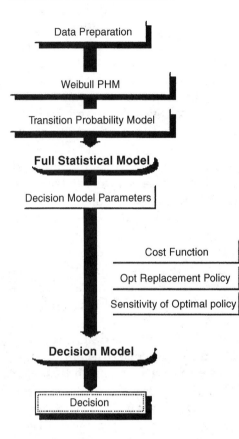

Figure 12-3 PHM flow diagram.

12.2 STEP 1: DATA PREPARATION

No matter what tools or computer programs are available, you should always examine the data in several ways (10). For example, many data sets can have the same mean and standard deviation and still be very different. That can be of critical significance. As well, instrument calibration and transcription mistakes have likely caused some data errors. You may have to search archives for significant data that's missing. To develop accurate decision models, you must be fully immersed in the operating and maintenance context. You must know the failure and repair Work Order process. Properly collected and validated subjective data that reflect all that is currently known about a problem is invaluable, but it's not enough. You also need to collect sound objective data about the problem or process, for a complete analysis and to confirm subjective opinion (11).

Generally, the maintenance engineer or technologist starts out with an underlying model based on the type of data, where the observations came from, and previous experience. After obtaining or converting the data into well-structured database tables, you must verify the model in stages before plunging ahead. This is where the tools of descriptive statistics—graphics (frequency histograms and cumulative frequency curves) and numbers (mean, median, variance, skewness)—are very useful. Some of these EXAKT™ methods are described in the following sections.

Although the use of powerful Computerized Maintenance Management Systems (CMMS) and Enterprise Asset Management (EAM) systems is growing, too little attention has been paid to data collection. Existing maintenance information-management database systems are underused. A clear relationship between accurately recorded component-age data and effective maintenance decisions has not been established. Tradespeople need to be educated about the value of such data, so that they meticulously record it when they replace failed components. Consistent with the principles of Total Productive Maintenance described in Chapter 8, maintenance and operational staff are the true custodians of the data and the models it creates.

12.2.1 Events and Inspections Data

Unlike simple Weibull analysis, described in Chapter 9 and applied in Chapter 11, PHM requires two types of equally important information: *event* data and *inspection* data. Three types of events, at a minimum, define a component's lifetime or history:

- The beginning (B) of the component life (the time of installation)
- The ending by failure (EF)
- The ending by suspension (ES), e.g., a preventive replacement

These events are identified in the events graph in Figure 12-4. Additional events should be included in the model if they directly influence the measured data. One such event is an oil change (designated by OC in Figure 12-4 and Table 12-1). Into the model, you should input that at each oil change some covariates—such as the wear metals—are expected to be reset to zero. This will prevent the model from being "fooled" by periodic decreases in wear metals. This point is illustrated in Figure 12-5.

Periodic tightening or recalibrating the machinery may have similar effects on measured values and should be accounted for in the model. The events graph for a particular unit in Example 1 is shown in Figure 12-4.

Example 1: Table 12-1 displays a partial data set from a fleet of four haul truck transmissions. You can contact the author for a complete data set in electronic form (4). You need such data to build a proportional hazard model. The inspections in this case are oil-analysis results (^{b} in the Event

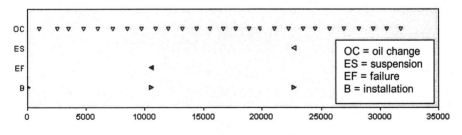

Figure 12-4 Events graph for unit HTT66.

Table 12-1 Haul Truck Transmission Inspection and Event Data

Ident	Date	Wage	HN	P	Event	Iron	Lead	Calcium	Mag
HTT66	12/30/93	0	1	4	B[a]	0	0	5000	0
HTT66	1/1/94	33	1	0	[b]	2	0	3759	0
HTT66	1/17/94	398	1	0	[b]	13	1	3822	0
HTT66	2/14/94	1028	1	0	[b]	11	0	3504	0
HTT66	2/14/94	1028	1	1	OC[c]	0	0	5000	0
HTT66	3/14/94	1674	1	0	[b]	10	0	4603	0
HTT66	4/12/94	2600	1	0	[b]	14	2	5067	0

[a] Beginning of life of component.
[b] Oil analysis results.
[c] Oil change.

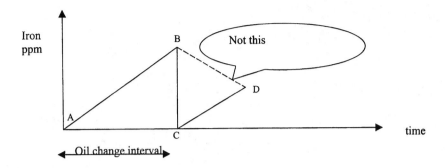

Figure 12-5 The actual transition is A-B-C-D, not A-B-D.

column. The data comprise the entire history of each unit, identified by the designations HTT66, HTT67, HTT76, and HTT77, between December 1993 and February 1998. The event and inspection data are displayed chronologically.

Example 2: In a food-processing company, shear pump bearings are monitored for vibration. In this example, 21 vibration covariates from the shear pump's inboard bearing (represented in Figure 12-6) are reduced to only the three significant ones shown in Table 12-2. As an exercise, you can develop and test the model using the same techniques described for Example 1 in the following sections.

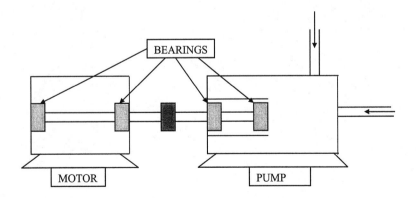

Figure 12-6 Shear pump.

Table 12-2   Significant Covariates

Ident.	Date	Working age	VEL_1A	VEL_2A	VEL_1V
B1	29-Sep-94	1	0.09	0.017	0.066
B1	08-Nov-94	41	0.203	0.018	0.113
B1	24-Nov-94	57	0.142	0.021	0.09
B1	25-Nov-94	58	0.37	0.054	0.074
B1	26-Nov-94	59	2.519	0.395	0.081
B2	11-Jan-95	46	0.635	0.668	0.05
B2	12-Jan-95	47	1.536	0.055	0.0078
B3	19-Apr-95	97	0.211	0.057	0.144
B3	04-May-95	112	0.088	0.022	0.079
B3	06-May-95	114	0.129	0.014	0.087
B3	29-May-95	137	0.225	0.021	0.023
B3	05-Jun-95	144	0.05	0.021	0.04
B3	20-Jul-95	189	1.088	0.211	0.318
B4	21-Jul-95	1	0.862	0.073	0.102
B4	22-Jul-95	2	0.148	0.153	0.038
B4	23-Jul-95	3	0.12	0.015	0.035
B4	24-Jul-95	4	0.065	0.021	0.018
B4	21-Aug-95	33	0.939	0.1	0.3

The bearing measurements were taken in three directions: axial, horizontal, and vertical. In each direction, the fast Fourier transform of the velocity vector was taken in five frequency bands. The overall velocity and acceleration were also measured. This provided a total of 21 vibration measurements from a single bearing. Example 2 analyzes these 21 signals using EXAKT™ software, concluding that only three of the vibration-monitoring measurements are key risk factors that need to be considered. They are two different velocity bands in the axial direction and one velocity band in the vertical direction.

By combining the risk model with economic factors, you can devise a replacement decision policy, such as the one represented by Figure 12-7. The cost of a failure repair, compared to a preventive repair, was input into the decision model, and the ratio of failure cost to preventive cost was 9:1. On the graph, the composite covariate Z—the weighted sum of the three significant influencing factors—are points plotted

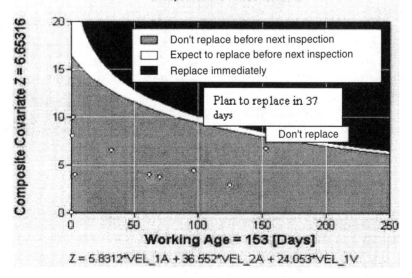

Figure 12-7 Optimal condition-based replacement policy.

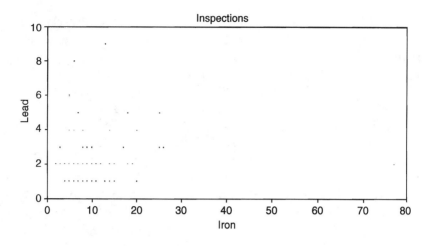

Figure 12-8   Lead versus iron cross graph.

against working age. If the current inspection point falls in the top section (above the curve), the unit should be repaired immediately, due to the combination of risk of failure and its cost exceeding an acceptable limit before the next inspection.

If the current point is in the bottom section (below the curve), the optimal decision to minimize the long-run cost is to continue operating the equipment and inspect it at the next scheduled inspection. If the point is in the light central zone, the optimal decision is to keep operating it but preventively replace the component before the next scheduled inspection. The chart also indicates how much longer (the remaining useful life) the equipment should run before being repaired or replaced.

Using the decision chart shown in Figure 12-7, total maintenance cost was reduced from $59.46/day to $26.83/day—an impressive 55% saving. Also, by following the recommendation on the chart, the mean time between bearing replacement is expected to increase by 10.2%.

12.2.2   Cross Graphs

The modeler must examine and completely understand the data in Table 12-1, and there are several software graphical tools that are a big help. The cross graph (Figure 12-8), for instance, simplifies graphical statistical analysis, such as the possible correlation between two diagnostic variables. This correlation becomes evident when the points are clustered around a straight line. If the points are randomly scattered, as they are in Figure 12-8 (which is a plot of lead versus iron), you can easily see that there is no correlation between the two covariates. If there is an obvious correlation, you can use this knowledge in subsequent modeling steps.

12.2.3 Events and Inspections Graphs

We highly recommend that you use various methods to validate the data. For example, errors in transcribing hour-meter readings are common. Figures 12-9 and 12-10 reveal just such a transcription error. The graph of working age versus calendar time shows an irregularity and, in the inspections table, the date and age (or usage) data are out of synchronization. This type of false data needs to be corrected before you build a decision model.

The technical term used by statisticians when the data contains inappropriate and misleading information is "dirty."

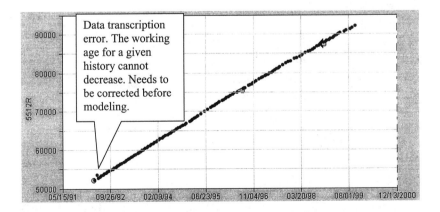

Figure 12-9 Working age versus calendar time showing data error.

Ident	Date	WorkingAge	All	Cr	Cu	Fe	
5512L	07/07/1999	91003	3	0	0	36	
Data Error – Note sequence	3/1999	91247	3	0	1	38	
	3/1999	91624	4	0	1	44	
	3/1999	92093	4	0	1	42	
	3/1992	52272	2	0	1	51	
		53534	1	0	1	31	
	1992	52796	2	0	1	40	
	0/1992	53086	2	1	1	51	
5512R	06/25/1992	53345	3	1	1	80	
5512R	07/12/1992	53610	1	0	0	48	
5512R	07/26/1992	53851	1	0	0	62	
5512R	08/14/1992	54106	1	0	0	33	
5512R	08/27/1992	54335	1	0	0	33	
5512R	09/12/1992	54509	1	0	0	14	

Figure 12-10 Data in inspections table to be corrected.

Dirty data must be cleaned before synthesizing it into a model for future decisions and policy.

The inspections graph should be used to investigate data outliers, which, if not valid, skew the information used to build the model. Instrument or transcription errors most frequently cause outliers. Eliminate these data points before building the model. Discuss the failure events (Figure 12-4) with the concerned maintenance and operational personnel. A failure from human error, such as wiring an electric motor in reverse, is normally included in the model not as a failure event but as a suspension.

If there are two or more independent failure modes, it may be best for the model to focus on the dominant one. In some cases, separate models are built for each kind of failure. In each model, you would classify the result of the other failure mode as a suspension. For a multicomponent system such as a diesel engine, it often makes sense to model major components such as the turbocompressor, fuel, and cooling systems separately.

12.2.4 Data Transformations

To model condition and age data, you need not only the actual information but any combinations (transformations) that are influential covariates. One obvious transformed data field of

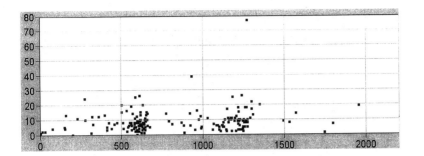

Figure 12-11 Iron versus oil age.

interest is the lubricating oil's age. This is usually not directly available in the database, but can be calculated knowing the dates of the oil changes (OC). In the current example, you would expect the "wear metals" iron or lead to be low just after an oil change and increase linearly as the oil ages and particles accumulate and remain in the oil circulating system. The cross graph in Figure 12-11 of oil age and iron negates this theory, except when the lubricating oil is relatively new. You can conclude that oil age, in this system, is *not* a significant covariate.

By applying a data "transformation" configurable in the software, you get new variables, shown in Table 12-3, such as oil age, component age (the column HW age), or the sum of two measurements (FeNi). Similarly, rates of change, rolling averages, and other combinations can be used in the premodeling statistical analysis (such as plotting various pairs of covariates in cross graphs) and in the model building.

12.3 STEP 2: BUILDING THE PHM

Examining the proportional hazards equation, you can see that it is an extension of the Weibull hazard function described in Chapter 9 and applied in Chapter 11.

$$h(t) = \frac{\beta}{\eta}\left(\frac{t}{\eta}\right)^{\beta-1} e^{\gamma_1 Z_1(t) + \gamma_2 Z_2(t) + \ldots + \gamma_n Z_n(t)} \tag{1}$$

Table 12-3 Inspections Including Transformed Data

Ident.	Date	Working age	HN	P	Event	HW age	Oil age	FeNi	Al	Cr	Cu	Fe	Ni
5501L	05/04/1991	40816	1	4	B	0	0	0	0	0	0	0	0
5501L	05/19/1991	41176	1	0	*	360	360	25	1	0	4	24	1
5501L	06/06/1991	41434	1	0	*	618	618	13	0	0	3	13	0
5501L	07/07/1991	41972	1	0	*	1156	1156	15	0	0	4	15	0
5501L	07/26/1991	42226	1	0	*	1410	1410	16	1	0	4	15	1
5501L	08/14/1991	42530	1	0	*	1714	1714	13	0	0	2	13	0
5501L	08/28/1991	42770	1	0	*	1954	1954	18	0	0	2	18	0
5501L	09/25/1991	43054	1	0	*	2238	2238	12	1	0	2	12	0
5501L	10/16/1991	43295	1	0	*	2479	2479	17	0	0	2	17	0
5501L	11/13/1991	43567	1	0	*	2751	2751	44	6	0	2	44	0
5501L	11/30/1991	43844	1	0	*	3028	3028	58	6	0	2	57	1
5501L	12/22/1991	44071	1	0	*	3255	3255	30	3	0	1	30	0
5501L	01/19/1992	44312	1	0	*	3496	3496	32	3	0	2	32	0
5501L	02/11/1992	44589	1	0	*	3773	3773	22	2	0	1	22	0
5501L	02/27/1992	44841	1	0	*	4025	4025	26	2	0	2	26	0

The new part factors in (as an exponential expression) the co-variates $Z_i(t)$. These are the measured signals at a given time t of, for example, the parts per million of iron or other wear metals in the oil sample. The covariate parameters γ_i specify the relative impact that each covariate has on the hazard function. A very low value for γ_1 indicates that the covariate isn't worth measuring. You'll find that software programs provide valuable criteria to omit unimportant covariates, as we cover in Sections 12.4, 12.5, and 12.6

To fit the proportional hazards model to the data, you must estimate not only the parameters β and η, as we did in the simple Weibull examples in Chapter 9, but also the covariate parameters γ_i.

Remember that condition monitoring isn't actually monitoring the equipment's condition per se, but variables that you *think* are related to it. Those variables, or covariates influence the failure probability shown by the hazard function $h(t)$ in Equation 1. From the model you construct, you want to learn each covariate's degree of influence (namely, the size of the covariate parameters γ_i), based on past data.

Although oil analysis is the inspection method used in Example 1, the covariates Z_i can be used in many ways. One or more covariates can come from the plant vibration monitoring program; for instance, $Z_i(t)$ could represent the vibration level at the first harmonic in the horizontal direction of an electric motor's inboard bearing. In vibration analysis, too, you have many data choices to help make decisions. Example 2 describes how, using historical data, 21 possible covariates, from readings in three directions and seven frequency bands, are reduced to the three main predictors of failure.

12.3.1 How Much Data?

In Chapter 9, we described the need to collect good data. That important message is a recurring theme in this chapter too. Model accuracy depends solely on the quantity and quality of the data collected and maintained. In Section 12.4 we describe various statistical tests to determine how confident you should be about the model.

Condition-monitoring data is quite often available in large quantities because most industrial plants today have regular predictive maintenance programs. On the other hand, lifetime data is usually lacking or inadequate. By that, we mean the dates when systems and components were replaced, repaired, or overhauled, when they failed, and how. Although collecting such data isn't simple, increased effort, awareness, and training will pay back significant future dividends.

The decision model's accuracy depends on how many histories have ended in failure and their corresponding condition data. Fleets of equipment, or groups of similar items such as bearings, obviously make good candidates for PHM.

12.3.2 Modeling the Data

After you have prepared and verified the data using the methods described in the previous sections, you're ready for the modeling work. The objective of this step is to estimate the parameters in the Weibull part of the proportional hazards model β and η, as well as the covariate parameters γ_i.

Table 12-4 shows that there are 138 lifetimes in the Example 1 data for modeling, of which 32 end in failure, 62 in preventive replacement, and 44 right-censored (they were still in service when the model was built). The percent of censored data helps you judge the model's validity. Model accuracy depends also on having as many component or system histories as possible ending in actual failure. Table 12-5 provides the parameter estimates for η, β, and the γ's (in the second column, Estimate).

Table 12-4 Summary of Events and Censored Values

Sample size	Failed	Censored (def.)	Censored (temp.)	% Censored
138	32	62	44	76.8

Table 12-5 Summary of Estimated Parameters (based on maximum-likelihood method) for Four Covariates

Parameter	Estimate	Sig.[a]	Standard error	Wald	DF	p-Value	Exp. of estimate	95% CI	
								Lower	Upper
Scale	3.318e+004		8840					1.585e+004	5.05e+004
Shape	1.755	Y	0.3087	5.983	1	0.01445		1.15	2.36
Sed.	8.638e-005	Y	2.097e-005	16.97	1	0	1	4.529e-005	0.0001275
Fe	0.002783	Y	0.0005467	25.91	1	0	1.003	0.001711	0.003854
Ni	-0.02071	N	0.05421	0.1459	1	0.7025	0.9795	-0.127	0.08554
Visc40	-0.003	N	0.002336	1.649	1	0.199	0.997	-0.00758	0.001579

[a] Based on 5% significance level.
Shape = 1 tested; gamma (cov.) = 0 tested.
See text for explanation of columns.

The software-assisted modeling process is iterative. You begin by choosing all the covariates you believe could be important influences on future failure. The program tests the significance of each covariate and reports the results, as in Table 12-5. Note that in this case, nickel (Ni) and viscosity (Visc40) are designated N—not significant. The next step is to eliminate one of them (the one with the highest *p*-value, nickel) from the covariate selection list and recalculate the PHM. You repeat the process, pruning the list, until only significant covariates are left.

Mathematically, there may not be a large enough number of lifetimes (machine histories ending in failure) to resolve all the model's simultaneous equations. This happens when you select too many covariates for the first model-building iteration. The program will advise you that there isn't enough historical data to resolve so large a model. Eliminating, one by one, the covariates that it indicates are insignificant will allow you to continue the process.

Table 12-5 illustrates a typical model estimation report. Note the N designation in the Significance column for nickel and viscosity at 40°C. The normal procedure at this point is to remove the nonsignificant covariates one at a time from the model and recalculate. In this example, through two iterations, we have eliminated nickel and viscosity. Table 12-6 is the report on the final model containing covariates iron (Fe) and sediment (Sed.). The parameter called "Shape" is also significant because the component's age influences its risk of failure. The shape parameter corresponds to the constant β of the Weibull distribution (discussed in Chapter 9).

The columns in Tables 12-5 and 12-6 are:

Estimate: lists the estimated scale, shape, and covariate parameters.

Sig.: Y or N. Indicates importance according to the Wald test—whether the shape parameter is close to 1 and the covariate parameters are close to 0. Y indicates

Table 12-6 Summary of Estimated Parameters (based on maximum-likelihood method) for Iron-Sediment Model

Parameter	Estimate	Sig.[a]	Standard error	Wald	DF	p-Value	Exp. of estimate	95% CI Lower	95% CI Upper
Scale	4.293e+004		8868					2.555e+004	6.031e+004
Shape	1.817	Y	0.3094	6.972	1	0.008279		1.211	2.423
Sed.	7.725e−005	Y	2.114e−005	13.35	1	0.0002587	1	3.581e−005	0.0001187
Fe	0.002673	Y	0.000291	84.36	1	0	1.003	0.002103	0.003244

[a] Based on 5% significance level.

that the shape is significantly different from 1 and the covariate parameters are significantly different from 0. A test significance level of 5% is applied, meaning that there is a 5% chance of being wrong that the significance is Y.

The remaining seven columns provide the Wald test calculation numbers that determine the Y or N in the significance column.

Standard error: shows how precise the estimates are. Larger standard deviations mean estimates are less precise. The standard error depends on the sample size and relationship between failure times and co-variates.

Wald: a test used to check whether risk can occur anytime and that the covariate parameter is close to 0. It checks whether the shape factor β is equal to 1, and if any of the covariate parameters γ are equal to 0. The test reports a probability value called a p-value. If the p-value is small (e.g., <5%–10%), the hypothesis ($\beta = 1$ or $\gamma = 0$) can be rejected (statistically).

DF: degrees of freedom refers to the chi-squared distribution used to calculate the p-value.

p-value: probability value as defined above in Wald.

Exp. of estimate: shows how much the hazard is related to this covariate. This is just another way to express (as a factor) the covariate's influence on the hazard.

95% CI: 95% confidence interval for each of the estimated parameters.

The maximum likelihood in the report heading refers to the estimation (curve-fitting) method used by the software.

12.4 STEP 3: TESTING THE PHM

At this point, you need to review the objectives of your stated maintenance management philosophy. Management's role is to establish policies and systems to garner the highest average return on investment in the long run, with "acceptable" risk.

In real life, when you estimate parameters from data, you can come up with many widely differing results, depending on the estimation method used. Due to the random nature of sample data, your arbitrary guess after looking at a few data points might be closer than the result offered by the most highly recommended computer program. Statistical theory describes how well various methods work compared to one another *in the long run*—over many, many applications—so it makes sense (from a long-term business perspective) to use the best long-run method on any single set of data you want to analyze (12). The following statistical testing methods are meant to provide "long-run" levels of confidence.

12.4.1 Residual Analysis

Residual analysis tells how well the PHM fits the data. These graphical techniques and statistical tests check whether some data points are not well enough represented by the estimated model (2). If the data contain numerous histories that end in suspensions (see Chapter 9 for an explanation of suspended data) or if the sample size (number of histories) is small, these procedures are not very accurate.

To test the model fit, you apply the Cox-generalized residuals method. Numbers called residuals ($r_1, r_2,...$) are calculated for every event time (either a failure or a suspension). Then you test whether the residuals $r_1, r_2,...$follow, statistically, the negative exponential distribution, as you expect if the model fits the data.

12.4.1.1 Graphical Techniques

There are four types of residual plots to look at the model fit. Each successive plot enhances visibility and your ability to judge the fit. The first plot is of residuals in order of appearance, Figure 12-12. This graphical method plots the residuals in the same order as the histories in the inspections table (Table 12-1). The average residual value must equal 1. So, expect

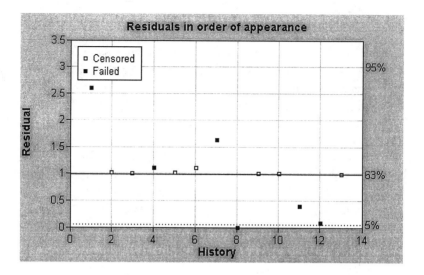

Figure 12-12 Residuals in order of appearance.

a random scatter of points around the horizontal line $y = 1$ if the model fits the data well. Note that the residuals from censored values are always above the line $y = 1$. To help in examining residuals, the 5% lower limit (LL = 0.05) and the 95% upper limit (UL = 0.95) are included on the graph. If the Weibull PHM fits data well, at least 90% of the residuals will be within these limits.

Another interesting plot is of ordered residuals against expectations, in Figure 12-13. This graphical method uses the residuals $r_1, r_2,...$arranged in ascending order against their expected values. If the Weibull PHM fits the data well, the points will be scattered around the line $y = x$. The difference between consecutive expectations increases, so the concentration of points in 50%–60% of the cases will be below the value 1. The residuals' variability also increases with order number. The points, then, on the right side of the plot needn't be close to the line $y = x$ for the model to be appropriate. You can adjust with a suitable transformation of residuals in the next plot, of transformed ordered residuals, Figure 12-14.

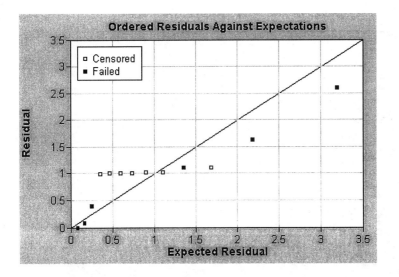

Figure 12-13 Ordered residuals against expectations.

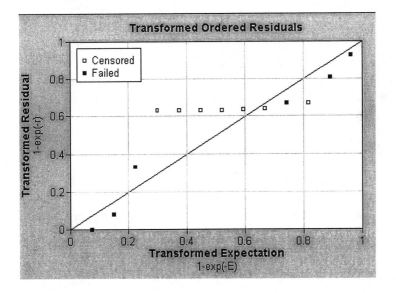

Figure 12-14 Ordered residuals transformed to displace all points between 0 and 1, and with approximately the same distance on the *x*-axis.

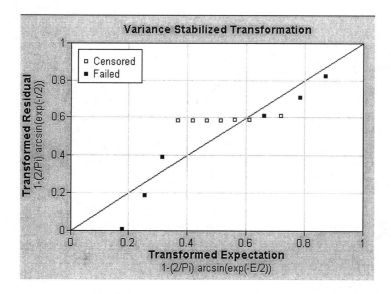

Figure 12-15   Variance-stabilized transformation.

The final plot, Figure 12-15, is of variance-stabilized transformation, which transforms the data to displace all points between 0 and 1. The variance has been "stabilized" to further reduce noise and improve ability to judge the fit.

12.4.1.2   The Kolmogorov-Smirnov (K-S) test

Similar to the Wald test discussed in Section 12.3.2 to determine whether component age and the covariates were significant, the K-S test indicates whether and how well the model fits the data. Table 12-7 displays a summary report of the PHM goodness-of-fit test (based on Cox-generalized residuals).

The K-S statistical test checks whether the residuals calculated from an estimated Weibull PHM follow the negative exponential distribution. The test calculates the distance between the theoretical exponential distribution and the

Table 12-7   PHM Goodness-of-Fit Test (based on Cox-generalized residuals)

Summary of events and censoring values

Sample size	Failed	Censored (def.)	Censored (temp.)	% Censored
13	6	3	4	53.8

Summary of estimated parameters

Scale	Shape	Iron	Lead
38994	5.00736	0.262649	1.05221

Summary of goodness-of-fit test

Test	Observed value	p-Value	PHM fits data
Kolmogorov-Smirnov	0.288364	0.191285	Not rejected[a]

[a] Based on 5% significance level.

distribution estimated from the residuals, adjusted for suspensions, and reports the p-value (probability value). This test will not work well if the suspensions are not randomly distributed. For instance, if a warranty limit requires that parts be replaced unilaterally at a fixed time (working age), this doesn't disqualify the model, but it's not an appropriate test of the model's accuracy.

The hypotheses in Table 12-7 that the PHM fits the data is not rejected. That conclusion comes from observing the p-value. If the p-value is small enough ($<5\%$–10%), the hypothesis should be rejected. If the p-value is large enough ($>5\%$–10%), the hypothesis can be accepted. That's the reason why the PHM-fits-data conclusion is not rejected. This is based on a significance level of 5%, i.e., 0.05. This test is not particularly reliable, though, with a large number of suspended values in the sample and/or a small sample size.

Table 12-8 PHM Parameter Estimation—Comparison

Submodel	Close to Base[a]	Deviance change	DF	Proba-bility	Parameter	Estimate	Standard error	Wald	Proba-bility
Base iron-lead		0	0	1	Scale	3.899e+004	1.638e+004		
					Shape	5.007	1.157	11.99	0.000535
					Iron	0.2626	0.07739	11.52	0.0006888
					Lead	1.052	0.376	7.83	0.005139
Iron	N	10.6114	1	0.00112	Scale	1.895e+004	3076		
					Shape	5.505	1.577	8.165	0.00427
					Iron	0.2292	0.05106	20.15	0

[a] Based on 5% significance level.

12.4.1.3 Comparative Analysis

By now, you will have examined and recorded a number of models, rejected some (nonsignificant covariates), and identified several whose fit is acceptable. You must then compare alternative models to see which is the "best." As a general rule, the least complex model (the one with the smallest number of covariates) that still fits the data is preferred. Using comparative analysis to compare the models, decide which one to use for the next phase—building the optimal decision policy model.

The comparison's main goal is to ascertain whether a simpler submodel can adequately replace a (bigger) parent model containing more covariates. The comparison uses a chi-squared test based on deviance change. The deviance is a value for every submodel during the estimation procedure. The difference between the basic submodel deviance and that of another submodel is called the *deviance change*, used to test the hypothesis that two submodels are statistically equivalent. For every deviance change, a p-value is calculated. For the basic parent model, the deviance change is equal to 0 and the p-value $= 1$ by definition, because the basic model is always compared to itself. If the p-value for a submodel is small (e.g., $<5\%$–10%), it may not be good enough to replace the basic one. In Table 12-8, the iron submodel is rejected for that reason. Comparing two submodels, the one with the higher p-value (column 5) is the one that better represents the data.

Columns 8, 9, and 10 provide additional insight. For example, column 8 puts confidence intervals on the parameter estimates of column 7.

12.5 STEPS 4 AND 5: THE TRANSITION PROBABILITY MODEL

12.5.1 Covariate Bands

You have developed and tested the Proportional Hazards Model using the software tools described in the previous sections. For each of the covariates in the model, you must now define ranges of values or states, for example, low, medium, high, or normal, marginal, and critical levels. These covariate bands are set by gut feel, plant experience, eyeballing the data, or a statistical method. This could be placing the boundary at two standard deviations above the mean to define the medium level and at three standard deviations to define the high value range.

12.5.2 Discussion of Transition Probability

Once you have established the covariate bands, the software can calculate the Markov chain model transition-probability matrix, Table 12-9, which shows the probability of going from one state (for example, light contamination, medium contamination, heavy contamination) to another, between inspection intervals.

Given a variable's present value range, you can predict (based on history) with a certain probability what it will be at the next inspection. For example, if the last inspection showed an iron level less than 13 parts per million (ppm) you can predict that the next record will be between 13 and 26 ppm with a probability of 12% (0.11918 in Table 12-9) and more than 26 ppm with a probability of 1.6%. Probabilities such as these are called *transition probabilities*.

12.5.3 Covariate Groups

Here is another way to make the model more accurate. Within the transition-probability model, you can group the covariates

Table 12-9 The Markov Chain Model Transition-Probability Matrix

Iron	0 to 13.013	13.013 to 25.949	Above 25.949
0 to 13.013	0.864448	0.11918	0.0163723
13.013 to 25.949	0.139931	0.657121	0.202947
Above 25.949	0	0	1

that you believe are related. Perhaps you found correlation or partial correlation in the premodeling exercises discussed in Section 12.2.2. If high values of one variable usually correspond with the same in another, they should be specified as a group of dependent variables for the transition-probability model. Variables from different groups are then assumed to be independent. The default is that all covariates are in the same group. In other words, they are all assumed dependent on one another. Grouping choices constitute a hypothesis that is, once again, statistically tested in the software as part of the modeling process. Table 12-10 is based on Example 2 data that assumes that the vibration in the axial direction of bearing A isn't related to the vibration in another direction and in a second band. That hypothesis is rejected.

A chi-squared test determines whether the assumed covariate grouping is appropriate, that is, whether variables from different groups are independent. The test calculates the

Table 12-10   Testing the Hypothesis that Two Groups of Condition Data Are Independent

Grouping				
Covariate				Group
VEL_1A				1
VEL_2A				0
VEL_1V				0

	Time intervals (days)			
	Above 0	Total	p-value	MCM fits data
Observed chi-squared	59.9546	59.9546		
DF+	27	27	0.00027	Rejected[a]
DF(Avg)	20.63	20.63	0.00001	Rejected[a]
DF−	6	6	0.00000	Rejected[a]

[a] Based on 5% significance level.

observed chi-squared value by comparing the transition frequencies for a given grouping, then reports a p-value. Selected
time intervals are also taken into account. The chi-squared
distribution depends on the parameter DF (degrees of freedom). The exact calculation of DF is complicated, so three in-
terpolated values of DF are used to calculate three versions
of the p-value: less conservative (DF+), more conservative
(DF−), and average (DF(Avg.)). How to test the hypothesis
that the covariate groups are independent: reject if DF+ p-
value is small (<5%–10%); accept if DF− p-value is large
(>5%–10%); use DF(Avg.) p-value in other cases. The test can
be used if there are at least two groups.

12.5.4 Time Intervals

You may suspect that different time periods in the component's life exhibit different transition probabilities. By speci-
fying those time periods in the model, you can improve model
accuracy. As in the determination of independent covariate
groups, there is a software test to determine whether the transition behavior among states is different for various time periods in the life of a machine or component. This function,
called the test for homogeneity, provides further insight into
the model.

You specify to the software that two or more time periods
may exhibit different transition behaviors. Then a chi-squared
test performed by the software determines whether the Markov chain is homogeneous, i.e., whether the transition probabilities depend on time (working age) periods you guessed. The
test calculates an observed chi-squared value, comparing the
transition rates calculated from the one time interval, and the
specified time intervals, and reports a p-value. If the test reports that indeed time periods are significant with regard to
covariate transitions, they should be left in the model. Otherwise they should be removed.

Table 12-11 shows the results of the homogeneity test for
the Example 1 data. In this case, the hypothesis of a wear-in

Table 12-11   Determining Whether Transition Probabilities Depend on Time

Time intervals (hr)	0–500	Above 500	Total	Max. time
Inspections	19 (7.4%)	238 (92.6%)	257	12639

	Time intervals (hr)				MCM fits data
	0–500	Above 500	Total	p-value	
Observed chi-Squared	12.0697	0.554869	12.6246		
DF+	0.5	4.5	5	0.02716	Rejected[a]
DF(Avg)	0.31	2.81	3.13	0.00623	Rejected[a]
DF−	0	0	0	Undefined	Undetermined

[a] Based on 5% significance level.

interval between 0 and 500 hours is not rejected. To test the hypothesis that transition probabilities don't depend on time intervals: reject the hypothesis if DF+ p-value is small (<5%–10%); accept it if DF− p-value is large (>5%–10%); use DF(Avg.) p-value in other cases, but with caution. This test can be used if at least two time intervals are included. It is not reliable if a small number of inspections occur in any interval.

12.6   STEP 6: THE OPTIMAL DECISION

12.6.1   The Cost Function

In Chapters 10 and 11, you learned how models optimize an objective such as the overall long-run cost to maintain a system. An analogous process is used in PHM. C designates the cost of a planned repair. An amount K is added to C if the repair occurs as a result of failure. K would include the extra costs due to the consequences of the item's having failed in service. The planned-to-failure cost ratio $C/(C + K)$ is blended into the model represented by the cost function graph, Figure 12-16, which shows the minimum cost ($0.358/hr) associated

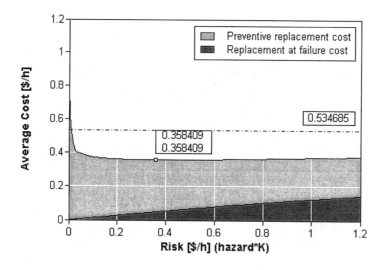

Figure 12-16   Cost function.

with the optimal replacement policy determined by the program.

The abscissa measures the risk. When the risk level goes
up (toward the right-hand edge of the graph), the cost of failure replacement and repair also increases. As the risk rises
to infinity, the average cost of maintenance approaches the
cost of a run-to-failure policy, $0.53/hr. The average long-run
cost of maintenance is the sum of the costs due to preventive
replacements (upper part zone) and failure replacements
(lower zone). You want to select the lowest possible risk level
(horizontal axis), without increasing the total maintenance
cost per hour. Naturally, zero risk would entail infinite cost.
An infinite risk is tantamount to a run-to-failure policy whose
cost is indicated by the dashed line.

12.6.2   The Optimal Replacement Decision Graph

The replacement decision graph, Figure 12-17, reflects and
culminates the entire modeling exercise to date. It combines the proportional hazards model results, the transition-

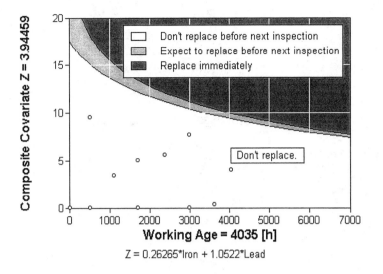

Figure 12-17 The optimal replacement decision graph.

probability model, the various hypotheses accepted for groups and time intervals, and the relative costs of failure and planned repairs into the best (lowest-cost) decision policy for the component or system in question.

The ordinate is the composite covariate Z, a balanced sum of covariates that statistically influence failure probability. Each covariate contributes to the risk of failure in the next inspection interval according to its influence as determined by the decision model.

A major advantage of this system is that a single graph combines the information you need to make a replacement decision. The alternative is to examine the trend graphs of perhaps dozens of parameters and guess whether to replace the component immediately or a little later. The optimal replacement decision graph recommendation is the most effective guide to minimize maintenance cost in the long run. It provides an unambiguous decision that can be automated by integrating the model into the maintenance management computer systems and associated business processes.

12.7   STEP 7: SENSITIVITY ANALYSIS

How do you know that the optimal replacement decision graph constitutes the best policy, considering your plant's ever-changing operations? Are the assumptions you used still valid and, if not, what will be the effect of those changes? Is your decision still optimal? These questions are addressed by sensitivity analysis.

The assumption you made in building the cost function model centered on the relative costs of a planned replacement versus those of a sudden failure. That cost ratio may have changed. If your accounting methods don't provide precise repair costs, you had to estimate them when building the cost portion of the decision model. In either case, these uncertain costs can create doubts about whether the optimal replacement decision graph policy is well founded.

The sensitivity analysis allays unwarranted fears, and indicates when to obtain more accurate cost data. Figure 12-18,

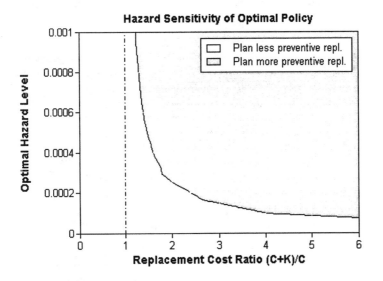

Figure 12-18   Sensitivity of optimal policy to cost ratio.

the hazard sensitivity of optimal policy graph, shows the relationship between the optimal hazard or risk level and the cost ratio. If the cost ratio is low, less than 3, the optimal hazard level would increase exponentially. We need, then, to track costs very closely to substantiate the benefits calculated by the model. On the other hand, if the cost ratio is in the 4 to 6 range, where the curve is fairly flat, changes in the cost ratio will not greatly affect the benefits derived from adhering to the optimal replacement policy determined by the model.

The cost sensitivity of optimal policy graph, Figure 12-19, has two lines. *Solid line*: If the actual (C + K)/C (the cost of a failure replacement divided by the cost of a preventive replacement) differs from that specified when you built the model, it means that the current policy (as dictated by the optimal replacement graph) is no longer optimal. The solid line tells what percentage you will be paying (when you adhere to the policy) above the optimal cost/unit time originally calculated. *Dashed line*: Again, assume the actual cost ratio has strayed. How much will the optimal average cost of mainte-

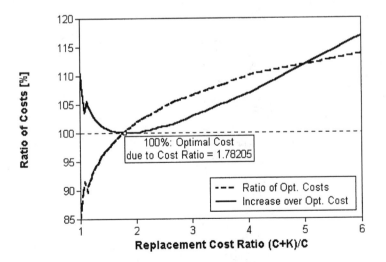

Figure 12-19   Cost sensitivity of optimal policy.

Table 12-12   Summary of Recorded Benefits

Industry	Data reduction of key condition indicators	% Cost savings over run-to-failure policy or simple age-based maintenance policy	Average extension in replacement life
Mining	21 oil-analysis measurements, 3 found to be significant	25	13%
Mass transit	A single-color observation used	55	More frequent maintenance inspections but less extensive repairs required.
Food processing	21 vibration signals, 3 found to be significant	5	
Petrochemical	12 vibration signals, 2 found to be significant	42	

nance change if you rebuilt the model using the new ratio of (C + K)/C. The dashed line tells how much your new optimal cost would change if you follow the new policy. The sensitivity graphs assume that only K (failure repair cost) changes and C (planned repair cost) remains the same.

12.8 CONCLUSION

In this chapter, we have explored a new approach to presenting, processing, and interpreting condition data. You have seen the benefits of applying proportional hazards models to condition-monitoring and equipment-performance data. Several industries, including petrochemicals, mining, food processing, and mass transit have successfully applied these techniques. Table 12-12 summarizes the advantages of CBM optimization by proportional hazards modeling in four companies. Good data and the increasing use of software will fuel even greater maintenance progress in the coming years.

REFERENCES

1. Trademark of Oliver Interactive Inc.
2. Brian D. Bunday. Statistical Methods in Reliability Theory and Practice. Upper Saddle River, NJ: Prentice Hall, 1991.
3. www.mie.utoronto.ca/labs/cbm.
4. murray.z.wiseman@ca.pwcglobal.co. Data sets for the exercises and software are available in electronic form from the author.
5. John Moubray. Reliability-Centered Maintenance. 2nd ed. Oxford: Butterworth-Heinemann, 1997.
6. K. Pottinger, A. Sutton. Maintenance management: an art or science? Maintenance Management International, 1983.
7. M. Anderson, A.K.S. Jardine, R.T. Higgins. The use of concomitant variables in reliability estimation. Modeling and Simulation 1982; 13:73–81.
8. F.S. Nowlan, H. Heap. Reliability-Centered Maintenance. Springfield, VA: National Technical Information Service, US Department of Commerce, 1978.

9. Murray Wiseman, A.K.S. Jardine. Proportional Hazard Modelling: A New Weapon in the CBM Arsenal. Proceedings of the Condition Monitoring '99 Conference, University of Wales, Swansea, April 12–15, 1999.

10. Paul A. Tobias, David C. Trinidade. Applied Reliability. 2nd ed. Boca Raton, FL: CRC Press, 1995.

11. Philip A. Scarf. On the application of mathematical models in maintenance. European Journal of Operational Research, 1997.

12. EXAKT Users' Manual. Condition Based Maintenance Laboratory, University of Toronto. www.mie.utoronto.ca/cbm.

13

Achieving Maintenance Excellence

A friend who travels extensively for his work in the mining industry describes his most anxious moment abroad: "I was just back in São Paulo from the Amazon basin, driving and not paying attention, when it dawned on me that I hadn't a clue where I was. It was dark, I don't speak Portuguese, my rental was the best car in sight, and I was sure I was getting eyeballed by some of the locals." Luckily, he got back to familiar territory, but not without a lot of stress and wasted effort. Like this intrepid traveler, you need to keep your wits about you to successfully reach your destination. So, to achieve maintenance excellence, you must begin by first checking where you are.

The preceding chapters described the evolution from reactive to proactive maintenance, managing equipment reliability to reduce failure frequency, and optimizing equipment performance by streamlining maintenance for total life-cycle economics. In this final chapter, we look at the specifics of implementing the concepts and methods of this process.

We take you through a three-step approach to put a maintenance excellence program into place and achieve results:

- **Step 1: Discover**. Learn where you are in a maintenance maturity profile, establish your vision and strategy based on research and benchmarking, and know your priorities, the size of the gap, and how much of it you want to close for now
- **Step 2: Develop**. Build the conceptual framework and detailed design, set your action plan and schedule to implement the design, obligate the financial resources, and commit the managers and skilled staff to execute the plan
- **Step 3: Deploy**. Document and delegate who is accountable and responsible, fix milestones, set performance measures and reporting, select pilot areas, and establish detailed specifications and policy for procurement, installation and training

Finally, we review the why's and how's of managing successful change in the work environment.

Before you implement reliability improvement and maintenance optimization, you must set up an organization or team mandated to effect change. You will need an executive sponsor to fund the resources and a champion to spearhead the program. You will need a steering group to set and modify direction. Members are typically representatives from the affected areas, such as maintenance, operations, materials, information technology, human resources, and engineering. A facilitator is invaluable, particularly one who has been through this process before and understands the shortcuts and pitfalls. Last, but certainly not least, you need a team of dedicated

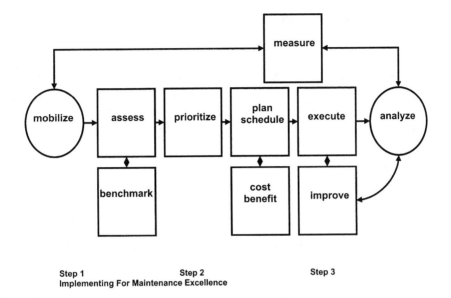

Figure 13-1 Steps to implementing for maintenance excellence.

workhorses to execute the initiatives. The process for achieving maintenance excellence is depicted in Figure 13-1.

13.1 STEP 1: DISCOVER

13.1.1 Mobilize

Mobilizing is getting off the mark—establishing the final project structure; staking out a work area, conference room, or office; and setting the broad "draft" project charter. You then develop the "first-cut" work plan, which will undoubtedly be modified after the next step, assessment.

13.1.2 Assess

This answers the "Where am I?" question, and follows a strict methodology to ensure completeness and objectivity.

13.1.2.1 Self-Diagnostic

The employees of the plant, facility, fleet, or operation give feedback on maintenance management proficiency—identifying areas of strength and weakness at a fairly high level, based on a standard questionnaire. One is shown in Appendix 24, designed as an initial improvement assessment of:

- Current maintenance strategy and acceptance level within the operations
- Maintenance organization structure
- Human resources and employee empowerment
- Use of maintenance tactics—PM, PdM, CBM, RTF, TBM, etc.
- Use of reliability engineering and reliability-based approaches to equipment performance monitoring and improvement
- Use of performance monitoring, measures, and benchmarking
- Use of information technology and management systems with particular focus on integrating with existing systems or any new systems needed to support best practices (e.g., document management, project planning)
- Use and effectiveness of planning and scheduling, and shutdown management
- Procurement and materials management that support maintenance operations
- Use of process analysis and redesign to optimize organizational effectiveness

The questionnaire can be developed with different emphases, depending on the area of focus. For example, for maintenance optimization, research could be conducted on what the Best Practice companies are doing with expert systems, modeling techniques, equipment and component replacement decision making, and life-cycle economics. Questions are posed to reflect how these Best Practices are understood and adopted. This self-assessment exercise builds ownership of the need to change, improve, and close the gaps between current and best practices. The questionnaire replies are summarized, graphed, and analyzed, and augment information from the next activity.

13.1.2.2 Data Collection and Analysis

While the questionnaire is being completed, gather operating-performance data at the plant for the review. The data needed includes:

- Published maintenance strategy, philosophy, goals, objectives, value, or other statements
- Organization charts and staffing levels for each division and its maintenance organization
- Maintenance budgets for the previous year (showing actual costs compared with budgeted costs, noting any extraordinary items) and for the current year
- Current maintenance-specific policies, practices, and procedures (including collective agreement, if applicable)
- Sample maintenance reports currently in use (weekly, monthly, etc.)
- Current process or workflow diagrams or charts
- Descriptions or contracts concerning outsourced or shared services
- Descriptions of decision support tools
- Summaries of typical spreadsheets, databases, and maintenance information-management systems
- Descriptions of models and special tools
- Position/job descriptions used for current maintenance positions, including planning, engineering, and other technical and administrative positions, as well as line functions

Although the above information is usually sufficient, you sometimes discover additional needs once work starts at the plant. Compare against the background data and reconcile any issues that come to light.

13.1.2.3 Site Visits and Interviews

If you're not familiar with the plant's layout or operation, you will need to spend time on a tour and learning safety procedures. A thorough tour that follows the production flow through the plant is best.

To facilitate the interview process on site, present the self-diagnostic results to management and other key personnel in a "kickoff" meeting. This will serve as an introduction to Best Practice. Introduce what you are doing and describe generic industrial Best Practices and what you will do on site. Then, conduct interviews of various plant personnel, using both the self-assessment and documentation collected earlier as question guides.

The interview questions are driven somewhat by areas of weakness and strength revealed through the self-diagnostic and through what you learn as you progress through the on-site diagnostic work. The interviewees will generally be:

- Plant manager, human resources/industrial relations manager
- Operations/production managers
- Information technology manager and systems administrators
- Purchasing manager, stores manager/supervisors
- Maintenance manager, superintendent, maintenance/ plant engineers, planners, supervisors
- Several members of the maintenance workforce (at least two from each major trade or area group)
- Representatives of any collective bargaining unit or employee association

Conduct interviews in either a private offices or the plant while the interviewee walks through his or her job and workplace. You may also want to observe how a planner, mechanic, or technician, for instance, spends the workday. This can reveal a lot about systemic and people issues, as well as opportunities that can impact Best Practices.

13.1.2.4 Maintenance Process Mapping

A group session is one of the best ways to identify major processes and activities. The processes should be the actual ones practiced at the plants, which may not coincide with the process maps developed for an ISO 9000 certification. Treat existing charts or maps as "as-designed" drawings, not necessarily "as-is" ones. These are mapped graphically to illustrate how work, inventory, and other maintenance practices are managed and performed. Through the mapping process, draw out criticisms of the various steps to reveal weaknesses. This not only will help you understand what is happening now but can be used for the "as-is" drawings. Or it can provide a baseline for process redesign, along with Best Practices determined from benchmarking or expert advice.

Through process mapping, you will gain significant insight into the current degree of system integration and areas where it may help. Redesigning processes will be part of any implementation work that follows the diagnostic.

The key processes to examine in your interviews and mapping include:

- PM development and refinement
- Procurement (stores, nonstores, services)
- Demand/corrective maintenance
- Emergency maintenance
- Maintenance prevention
- Work Order management
- Planning and scheduling of shutdowns
- Parts inventory management (receiving, stocking, issuing, distribution, review of inventory investment)
- Long-term maintenance planning and budgeting
- Preventive/predictive maintenance planning, scheduling, and execution

Through the interviews, identify and map other processes unique to your operations.

13.1.2.5   Report and Recommendations

At the end of the site visit, compile all the results into a single report showing overall strengths and weaknesses, the specific performance measures that identify and verify performance, and the most significant gaps between current practices and the vision. The report should contain an "opportunity map" that plots each recommendation on a grid, showing relative benefit compared to the level of difficulty to achieve it (see Section 13.2.1).

13.1.3   Benchmark

In many companies, benchmarking is more industrial tourism than an improvement strategy. After you have completed a diagnostic assessment, select the key factors for maintenance success specific to your circumstances. You need to focus sharply to gain useful information that can actually be implemented. Look both within and outside your industry to discover who excels at those factors. Compare performance measures, the process they now use, and how they were able to achieve excellence. But finding organizations that have excellent maintenance optimization can be difficult. There may only be a few dozen, for example, who have successfully implemented condition-based maintenance optimization using stochastic techniques and software.

13.2   STEP TWO: DEVELOP

13.2.1   Prioritize

You will not be able to implement everything at once. In fact, limited resources and varying benefits will whittle down your "short list." One technique to quickly see the highest-value initiatives versus those with limited benefit and expensive price tags is shown in Figure 13-2.

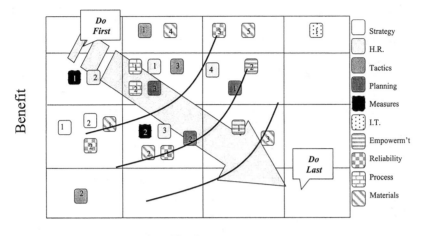

Benefit

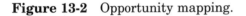

Degree of Difficulty

Figure 13-2 Opportunity mapping.

The benefit is often measured in a log 10 savings scale, e.g., $10,000, $100,000, $1,000,000, or $10,000,000. The degree of difficulty could be shown on a similar scale, an implementation timeline such as 1 year, 2 years, 3 years, 4 years, or a factor related to the degree of change required in the organization for the initiative.

13.2.2 Plan and Schedule

Using software, such as Microsoft's MS Project, you will make the planning and scheduling process more rigorous. List the projects in priority (Section 13.2.1) and with credible timelines. No one will commend you for taking the fast track in the planning phase if the project fails several months along. Be sure to include people from other functions or processes related to maintenance, if they can help ensure project success. Typically, you'll look to procurement, production planning, human resources, finance and accounting, general engineering, information technology, contractors, vendors, and/or service providers.

13.2.3 Cost-Benefit

The priority assessment from Section 13.2.1 broadly groups the recommendations. If there is a significant capital or employee time investment, calculate the cost-benefit. Whereas costs are usually fairly straightforward—hardware, software, training, consulting, and time—estimating benefits can be a lot more nebulous. What are the benefits of implementing Root Cause Failure Analysis, an RCM program, or the EXAKT™ software for CBM optimization? In Chapter 1, we described how to estimate the benefits of moving to more planned and preventive maintenance, which can be modified for specific projects targeting unplanned maintenance. Often, the equipment and hardware manufacturer or software/methodology designer can help, based on their experience with similar applications. Benchmarking partners or Web sites focused on maintenance management are other sources of useful information.

13.3 STEP 3: DEPLOY

13.3.1 Execute

If possible, consider a pilot approach to improvement initiatives. Not only does this provide proof of concept; more importantly, it acts as an excellent marketing tool for a full rollout.

What does experience teach us about program management? Program management is managing a group of projects with a common theme. Some of its key elements are:

- Define the scope (one large enough to capture management's attention and get meaningful results)
- Follow a documented approach
- Delineate roles and responsibilities
- Put a lot of effort up front in the discover and develop phases (the measure-twice/cut-once approach)
- Assess the "doability" of the plan with all stakeholders
- Avoid shortcuts or fast-tracking

- Estimate the risk beyond the budget
- Work to get the right champion who will be in it for the long run

13.3.2 Measure

The more detailed the implementation plan, the easier it is to measure progress. Include scope, detailed activities, responsibilities, resources, budgets, timelines for expenditures and activities, and milestones. Ensure maximum visibility of these measures by posting them where everyone involved can see them. Often forgotten are the review mechanisms: supervisor to subordinate, internal peer group or project team, management reviews, start-of-day huddle. They are all performance measures.

13.3.3 Analyze/Improve

After the initial round of piloting and implementation comes the reality. Is the program actually delivering the expected results? How are the cost-benefit actuals measuring up? Despite best efforts and excellent planning and execution, there often still have to be adjustments to the original scope, work plan, staffing, or expected results. But if you have closely followed the discover and design stages, and set up and managed performance measures, there will be few surprises at this stage.

13.3.4 Managing the Change

Finally, let's look at managing successful change during this three-stage process. There are six absolutely critical elements:

- Ensure that everyone understands the compelling need to change the current order, to close the gap between what is done today and the vision.
- Build this vision of what the new order will be like, so that it is shared and all can buy into it.

- Obtain management commitment, staying power, and the financial resources to get the job done.
- Get buy-in from those most affected by the change, by linking performance with rewards and recognition.
- Monitor performance and exercise leadership and control, especially when the implementation begins to drift off course.
- Communicate results at every step of the way, and at regular intervals. This is the single most talked-about pitfall by organizations that have stumbled in achieving the results hoped for at the outset of an improvement project.

Every implementation to improve the way we manage physical assets will be different. This is because we are all individuals, operating in a unique company culture, implementing projects that are people-driven. In this book, we have tried to impart our knowledge of what works best in most cases, for you to take and apply to your own particular circumstances.

Although you will strive to use the most cost-effective methods and the best tools, these alone will not guarantee success. What will is a committed sponsor with sufficient resources, an enthusiastic champion to lead the way, a project manager to stay the course, and a motivated team to take action and set the example for the rest of the organization.

Appendix 1: Mean Time to Failure

The expected life, or the expected time during which a component will perform successfully, is defined as

$$E(t) = \int_0^\infty tf(t)dt$$

$E(t)$ is also known as the Mean Time to Failure, or MTTF.

Integrate $\int_0^\infty R(t)dt$ by parts using $\int udv = uv - \int vdu$ and letting $u = R(t)$ and $dv = dt$.

$$\frac{du}{dt} = \frac{dR(t)}{dt}$$

but by examining Figure 9-2 (the Probability Density Function) we see that

$$R(t) = 1 - F(t)$$

and

$$\frac{dF(t)}{dt} = f(t)$$

Therefore

$$\frac{du}{dt} = \frac{dR(t)}{dt} = -\frac{dF(t)}{dt} = -f(t)$$

Then $du = -f(t)dt$.

Substituting into $\int u\, dv = uv - \int v\, du$

$$\int_0^\infty R(t)dt = [tR(t)]_0^\infty + \int_0^\infty tf(t)dt$$

but $\lim_{t\to\infty} tR(t) = 0$ when $\int_0^\infty tf(t)dt$ exists. Therefore the term $[tR(t)]_0^\infty$ in the above equation is 0.

Hence

$$\int_0^\infty R(t)dt = \int_0^\infty tf(t)dt = E(t) = \text{MTTF}$$

Appendix 2: Median Ranks

When only a few failure observations are available (say, ≤12), use is made of median rank tables (Table A.2.1).

Example: Bearing failures times (in months): 2, 3, 3.5, 4, 6. From median rank tables:

	Median rank (%)	
First failure time	13.0	2 months
Second failure time	31.5	3 months
Third failure time	50.0	3.5 months
Fourth failure time	68.6	4 months
Fifth failure time	87.1	6 months

Table A.2.1 Median Ranks

	1	2	3	4	5	6	7	8	9	10	11	12
1	50	29.289	20.630	15.910	12.945	10.910	9.428	8.300	7.412	6.697	6.107	5.613
2		70.711	50.000	38.573	31.381	26.445	22.849	20.113	17.962	16.226	14.796	13.598
3			79.370	61.427	50.000	42.141	36.412	32.052	28.624	25.857	23.578	21.669
4				84.090	68.619	57.859	50.000	44.015	39.308	35.510	32.390	29.758
5					87.055	73.555	63.588	55.984	50.000	45.169	41.189	37.853
6						89.090	77.151	67.948	60.691	54.811	50.000	45.951
7							90.572	79.887	71.376	64.490	58.811	54.049
8								91.700	82.018	74.142	67.620	62.147
9									92.587	83.774	76.421	70.242
10										93.303	85.204	78.331
11											93.893	86.402
12												94.387

From a Weibull analysis:

$\mu = 3.75$ months

$\sigma = 1.5$ months

Benard's formula is a convenient and reasonable estimate for the median ranks.

$$\frac{\text{Cumulative probability estimator}}{\text{Benard's formula}} = \frac{i - 0.3}{N + 0.4}$$

Appendix 3: Censored Data

Hours	Event	Order	Modified order	Median rank
67	F	1	1	0.13
120	S	2	—	—
130	F	3	2.5	0.41
220	F	4	3.75	0.64
290	F	5	5.625	0.99

Procedure:

$$I = \frac{(n + 1) \times \text{Previous order number}}{1 + \text{Number of items following suspended set}}$$

where I = increment. The first-order number remains unchanged.

For the second, applying the equation for the increment I, we get

$$I = \frac{(5 + 1) \times (1)}{1 + (3)} = 1.5$$

Adding 1.5 to the previous order number 1 gives the order number of 2.5 to the second failure:

$$\text{Cumulative probability estimator} = \frac{i - 0.3}{N + 0.4}$$
$$\text{Benard's formula}$$

$$I = \frac{(5 + 1) \times (2.5)}{1 + (3)} = 3.75$$

$$I = \frac{(5 + 1) \times (3.75)}{1 + (3)} = 5.625$$

Applying Benard's Formula to estimate median ranks for the first failure:

$$\text{Median rank} = \frac{1 - 0.3}{5 + 0.4} = .13$$

Similarly, for the second, third, and fourth failures, respectively:

$$\frac{2.5 - 0.3}{5.4} = 0.41; \quad \frac{3.75 - 0.3}{5.4} = 0.64; \quad \frac{5.625 - 0.3}{5.4} = 0.99.$$

Appendix 4: The Three-Parameter Weibull Function

Failure no. i	Time of failure t_i	Median ranks $N = 20$ $F(t_i)$
1	550	3.406
2	720	8.251
3	880	13.147
4	1020	18.055
5	1180	22.967
6	1330	27.880
7	1490	32.975
8	1610	37.710
9	1750	42.626
10	1920	47.542
11	2150	52.458
12	2325	57.374
13–20	Censored data	

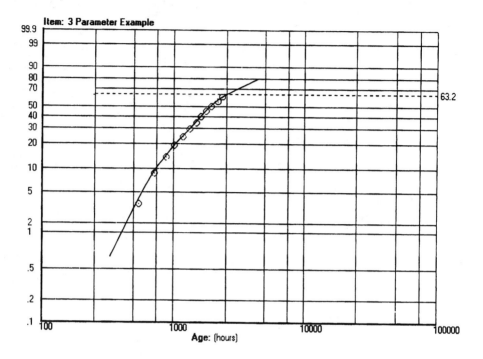

The curvature suggests that the location parameter is greater than 0.

Failure no. i	Time of failure t_i	Median ranks $N = 20$ $F(t_i)$
1	0	3.406
2	170	8.251
3	330	13.147
4	470	18.055
5	630	22.967
6	780	27.880
7	940	32.975
8	1060	37.710
9	1200	42.626
10	1370	47.542
11	1600	52.548
12	1775	57.374
13–20	Censored data	

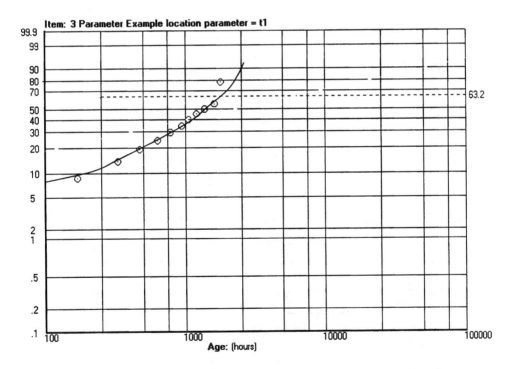

Now we get a line that is curved the other way, proving that the location parameter has a value between 0 and 550. $\lambda = 375$ yields a straight line, as shown in the following figure.

Failure no. i	Time of failure t_i	Median ranks $N = 20$ $F(t_i)$
1	375	3.406
2	495	8.251
3	705	13.147
4	845	18.055
5	1005	22.967
6	1155	27.880
7	1315	32.975
8	1435	37.710
9	1575	42.626
10	1745	47.542
11	1975	52.458
12	2150	57.374
13–20	Censored data	

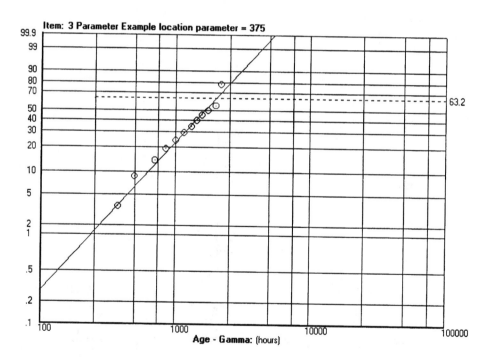

The three-parameter Weibull $\lambda = 375$.

Appendix 5: Confidence Intervals

From the Weibull plot, we can say that at time $t = 7$ the cumulative distribution function will have a value between 30% and 73% with 90% confidence. In other words, after 7 hours in 90% of the tests, between 30% and 73% of the batteries will have stopped working.

If we want a confidence interval of 90% on the reliability $R(t)$ at the time $t = 7$, we take the complement of the limits on the confidence interval for $F(t)$: (100–73, 100–30). So the 90% confidence interval for the reliability at time $t = 7$ hours is between 27% and 70%. Or we can say that we are 95% sure that the reliability after 7 hours will not be less than 27%.

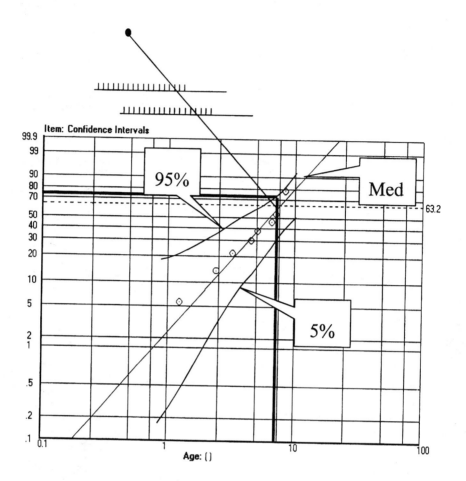

Table A.5.1   12 Electrical Batteries

Failure no.	Median ranks	5% ranks	95% ranks	t_i
1	5.613	0.426	22.092	1.25
2	13.598	3.046	33.868	2.40
3	21.669	7.187	43.811	3.20
4	29.758	12.285	52.733	4.50
5	37.853	18.102	60.914	5.00
6	45.941	24.530	68.476	6.50
7	54.049	31.524	75.470	7.00
8	62.147	39.086	81.898	8.25
9–12	Still operating			

Table A.5.2 5% Ranks

	1	2	3	4	5	6	7	8	9	10	11	12
1	5.00	2.53	1.70	1.27	1.02	0.85	0.71	0.64	0.57	0.51	0.47	0.43
2		22.36	13.054	9.76	7.64	6.28	5.34	4.62	4.10	3.68	3.33	3.05
3			36.84	24.86	18.92	15.31	12.88	11.11	9.78	8.73	7.88	7.19
4				47.24	34.26	27.13	22.53	19.29	16.88	15.00	13.51	12.29
5					54.93	41.82	34.13	28.92	25.14	22.24	19.96	18.10
6						60.70	47.91	40.03	34.49	30.35	27.12	24.53
7							65.18	52.9	45.04	39.34	34.98	31.52
8								68.77	57.09	49.31	43.56	39.09
9									71.69	60.58	52.99	47.27
10										74.11	63.56	56.19
11											76.16	66.13
12												77.91

Table A.5.3   95% Ranks

	1	2	3	4	5	6	7	8	9	10	11	12
1	95	77.64	63.16	52.71	45.07	39.30	34.82	31.23	28.31	25.89	23.84	22.09
2		97.47	86.46	75.14	65.74	58.18	52.07	47.07	42.91	39.42	36.44	33.87
3			98.31	90.24	81.08	72.87	65.87	59.97	54.96	50.69	47.01	43.81
4				98.73	92.36	84.68	77.47	71.08	65.51	60.66	56.44	52.73
5					98.98	93.72	87.12	80.71	74.86	69.65	65.02	60.90
6						99.15	94.66	88.89	83.13	77.76	72.88	68.48
7							99.27	95.36	90.23	85.00	80.04	75.47
8								99.36	95.90	91.27	86.49	81.90
9									99.43	96.32	92.12	87.22
10										99.49	96.67	92.81
11											99.54	96.95
12												99.57

Appendix 6: Estimating (Fitting) the Distribution

Maximum-likelihood estimate method:

$$\frac{\sum\limits_{i=1}^{n} x_i^{\hat{\beta}} \ln x_i}{\sum\limits_{i=1}^{n} x_i^{\hat{\beta}}} - \frac{1}{n} \sum\limits_{i=1}^{n} \ln x_i - \frac{1}{\hat{\beta}} = 0, \qquad \hat{\eta} = \left(\frac{\sum\limits_{i=1}^{n} x_i^{\hat{\beta}}}{n} \right)^{1/\hat{\beta}}$$

Method of least-squares estimate:

$$\hat{\beta} = \frac{\sum\limits_{i=1}^{n} x_i^2 - \dfrac{\left(\sum\limits_{i=1}^{n} x_i\right)^2}{n}}{\sum\limits_{i=1}^{n} x_i y_i - \dfrac{\sum\limits_{i=1}^{n} x_i \sum\limits_{i=1}^{n} y_i}{n}}, \qquad \hat{\eta} = e^{\hat{A}}$$

where

$$y_i = \ln(t_i) \text{ and } x_i = \ln[\ln(1 - \text{Median rank of } y_i)]$$

$$\hat{A} = \frac{\Sigma y_i}{n} - \frac{1}{\beta}\frac{\Sigma x_i}{n}$$

The MLE method needs an iterative solution for the beta estimate. Statisticians prefer maximum-likelihood estimates to all other methods because MLE has excellent statistical properties. They recommend MLE as the primary method. In contrast, most engineers recommend the least-squares method. In general, both methods should be used because each has advantages and disadvantages in different situations. Although MLE is more precise, for small samples it will be more biased than rank-regression estimates (from Ref. 2 in Chapter 9).

Appendix 7: Kolmogorov-Smirnov Goodness-of-Fit Test

STEPS

1. Determine the distribution to which you want to fit the data. Then determine the parameters of the chosen distribution.
2. Determine the significance level of the test (α usually at 1, 5 or 10%). It is the probability of rejecting the hypothesis that the data follows the chosen distribution assuming the hypothesis is true.
3. Determine $F(t_i)$ using the parameters assumed in step 1. $F(t_i)$ is the value of the theoretical distribution for failure number i.
4. From the failure data compute $\hat{F}(t_i)$ using the median ranks if appropriate.

5. Determine the maximum value of:

$$\left\{ \begin{array}{l} |F(t_i) - \hat{F}(t_i)| \\ |F(t_i) - \hat{F}(t_{i-1})| \end{array} \right\} = d$$

$d > d_\alpha$, we reject the hypothesis that the data can be adjusted to the distribution chosen in step 1 (d_α is obtained for the K-S statistic table).

We have tested five items to failure. Here are the failure times: 1, 5, 6, 8, and 10 hours. We assumed that the data follows a normal distribution and will check this assumption with a K-S goodness-of-fit test.

SOLUTION

Estimate the parameters of the chosen distribution: Estimate of

$$\mu = \frac{\Sigma t_i}{n} = 6$$

and the estimate of

$$\sigma^2 = \frac{\Sigma(t_i - t)^2}{(n - 1)} = s^2$$

t_i	$\hat{F}(t_i)$	$|F(t_i)|$	$|F(t_i) - \hat{F}(t_i)|$	$|F(t_i) - \hat{F}(t_{i-1})|$	d
1	0.070	0.129	0.059		0.059
5	0.390	0.314	0.076	0.261 d_{max}	0.261
6	0.500	0.500	0.0	0.186	0.186
8	0.720	0.686	0.034	0.220	0.220
10	0.880	0.871	0.009	0.194	0.194

The values of $F(t_i)$ are obtained from the normal distribution table. Here we are engaging in hypothesis testing. A significance level is applied by some authority or standard governing the situation, either 0.01, 0.05, 0.10, 0.15, or 0.20.

Table A.7.1   Kolmogorov-Smirnov Critical Values

Sample size n	K-S level of significance (d_α)				
	0.20	0.15	0.10	0.05	0.01
1	0.900	0.925	0.950	0.975	0.995
2	0.684	0.726	0.776	0.842	0.929
3	0.565	0.597	0.642	0.708	0.828
4	0.494	0.525	0.564	0.624	0.783
5	0.446	0.474	0.510	0.565	0.669
6	0.410	0.436	0.470	0.521	0.618
7	0.381	0.405	0.438	0.486	0.577
8	0.358	0.381	0.411	0.457	0.543
9	0.339	0.360	0.388	0.432	0.514
10	0.322	0.342	0.368	0.410	0.490
11	0.307	0.326	0.352	0.391	0.468
12	0.285	0.313	0.338	0.375	0.450
13	0.284	0.302	0.325	0.361	0.433
14	0.274	0.292	0.314	0.349	0.418
15	0.266	0.283	0.304	0.338	0.404
16	0.258	0.274	0.295	0.328	0.392
17	0.250	0.266	0.286	0.318	0.381
18	0.244	0.259	0.278	0.309	0.371
19	0.237	0.252	0.272	0.301	0.363
20	0.231	0.246	0.264	0.294	0.356
25	0.21	0.22	0.24	0.27	0.32
30	0.19	0.20	0.22	0.24	0.29
35	0.18	0.19	0.21	0.023	0.27
Over 35	$\dfrac{1.07}{\sqrt{n}}$	$\dfrac{1.14}{\sqrt{n}}$	$\dfrac{1.22}{\sqrt{n}}$	$\dfrac{1.36}{\sqrt{n}}$	$\dfrac{1.63}{\sqrt{n}}$

Source: Journal of the American Statistical Association 46(53):70.

We apply our K-S statistic, 0.261, to the row for a sample size of 5 of Table A.7.1. Assuming we wish to conform to a significance level of 0.20, we note that 0.261 is not greater than 0.446. That means that if we were to reject the model, there will be a high probability (20%) that we're wrong (in rejecting a good model). We therefore say that the model is not rejected based on a 20% significance level. Frequently the less stringent 5% significance level is applied.

Appendix 8: Present Value

PRESENT VALUE FORMULAE

To introduce the present-value criterion (or present discounted criterion) consider the following. If a sum of money—say, $1000—is deposited in a bank where the compound interest rate on such deposits is 10% per annum, payable annually, then after 1 year there will be $1100 in the account. If this $1100 is left in the account for a further year, there will then be $1210 in the account.

In symbol notation we are saying that if $L is invested and the relevant interest rate is i% per annum, payable annually, then after n years the sum S resulting from the initial invesment is

$$S = \$L\left(1 + \frac{i}{100}\right)^n \tag{1}$$

Thus, if $L = \$1000$, $i = 10\%$, and $n = 2$ years

$$S = 1000(1 + 0.1)^2 = \$1210$$

The present-day value of a sum of money to be spent or received in the future is obtained by doing the reverse calculation of that above. Namely, if $\$S$ is to be spent or received n years in the future, and $i\%$ is the relevant interest rate, then the present value of $\$S$ is

$$PV = \$S\left(\frac{1}{1 + \dfrac{i}{100}}\right)^n \tag{2}$$

where

$$\left(\frac{1}{1 + \dfrac{i}{100}}\right) = r \text{ is termed the discount factor.}$$

Thus, the present-day value of $\$1210$ to be received 2 years from now is

$$PV = 1210\left(\frac{1}{1 + 0.1}\right)^2 = \$1000$$

That is, $\$1000$ today is "equivalent" to $\$1210$ 2 years from now when $i = 10\%$.

It has been assumed that the interest is paid once per year. Interest may, in fact, be paid weekly, monthly, quarterly, semiannually, etc., in which case the formulae (1) and (2) need to be modified. In practice, with replacement problems it is usual to assume that interest is payable once per year and so Equation 2 is used in the present-value calculations.

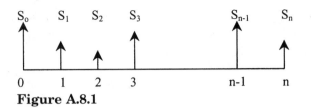

Figure A.8.1

It is usual to assume that the interest rate i is given as a decimal, not in percentage terms. Equation 2 is then written as

$$PV = \$S\left(\frac{1}{1+i}\right)^n \tag{3}$$

An illustration of the sort of problem in which the present-value criterion is used is the following. If a series of payments $S_0, S_1, S_2 \ldots S_n$, illustrated in Figure A.8.1, are to be made annually over a period of n years, then the present value of such a series is

$$PV = S_0 + S_1\left(\frac{1}{1+i}\right)^1 + S_2\left(\frac{1}{1+i}\right)^2 + \cdots + S_n\left(\frac{1}{1+i}\right)^n \tag{4}$$

If the payment is S_j, where $j = 0, 1, \ldots, n$ are equal, then the series of payments is termed an annuity and Equation 4 becomes

$$PV = S + S\left(\frac{1}{1+i}\right)^1 + S\left(\frac{1}{1+i}\right)^2 + \cdots + S\left(\frac{1}{1+i}\right)^n \tag{5}$$

which is a geometric progression and the sum of $n + 1$ terms of a geometric progression gives

$$PV = S\left[\frac{1 - \left(\dfrac{1}{1+i}\right)^{n+1}}{1 - \left(\dfrac{1}{1+i}\right)}\right] = S\left(\frac{1 - r^{n+1}}{1 - r}\right) \tag{6}$$

If the series of payments of Equation 5 is assumed to continue over an infinite period of time, i.e., $n \rightarrow \infty$, then from the sum to infinity of a geometric progression we get

$$PV = \frac{S}{1 - r} \tag{7}$$

In all the above formulae we have assumed that i remains constant over time. If this is not a reasonable assumption, then Equation 4 needs to be modified slightly; for example, we may let i take the values $i_1, i_2, \ldots$, in the different periods.

EXAMPLE: ONE-SHOT DECISION

To illustrate the application of the present-value criterion in order to decide on the best of a set of alternative investment opportunities we will consider the following problem.

A subcontractor obtains a contract to maintain specialized equipment for a period of 3 years, with no possibility of an extension of this period. To cope with the work, the contractor must purchase a special-purpose machine tool. Given the costs and salvage values shown in the table for three equally effective machine tools, which should the contractor purchase? We will assume that the appropriate discount factor is 0.9 and that operating costs are paid at the end of the year in which they are incurred.

Machine tool	Purchase price ($)	Installation cost ($)	Operating cost $			Salvage value
			Year 1	Year 2	Year 3	
A	50,000	1000	1000	1000	1000	30,000
B	30,000	1000	2000	3000	4000	15,000
C	60,000	1000	500	800	1000	35,000

For machine tool A:

Present value $= 50,000 + 1000 + 1000(0.9)$
$\qquad\qquad + 1000(0.9)^2 + 1000(0.9)^3 - 30,000(0.9)^3$
$\qquad\qquad = \$31,570$

For machine tool B:

Present value $= 30{,}000 + 1000 + 2000(0.9)$
$$+\ 3000(0.9)^2 + 4000(0.9)^3 - 15{,}000(0.9)^3$$
$$=\ \$27{,}210$$

For machine tool C:

Present value $= 60{,}000 + 1000 + 500(0.9)$
$$+\ 800(0.9)^2 + 1000(0.9)^3 - 35{,}000(0.9)^3$$
$$=\ \$37{,}310$$

Thus equipment B should be purchased since it gives the minimum present value of the costs.

FURTHER COMMENTS

In the above machine-tool-purchasing example, it will be noticed that the same decision regarding the tool to purchase would not have been reached if no account had been taken of the time value of money. Note also that many of the figures used in such an analysis will be estimates of future costs or returns. When there is uncertainty about any such estimates, or when the present-value calculation indicates several equally acceptable alternatives (because their present values are more or less the same), then a sensitivity analysis of some of the estimates may provide information to enable an "obvious" decision to be made. If this is not the case, then we may impute other factors such as knowledge of supplier, spares availability, etc., to assist in reaching a decision. Of course, when estimating future costs and returns, possible increases in materials costs, wages, etc. (i.e., inflationary effects), should be considered.

When dealing with capital-investment decisions, a criterion other than present value is sometimes used. For discussion of such criteria—for example, payback period and internal rate of return—the reader is referred to the engineering economic literature, such as Refs. 2 and 3 of the Bibliography in Chapter 9.

Appendix 9: Cost of Capital Required for Economic-Life Calculations

Assume that the cost associated with burrowing money is an interest charge of about 20% per annum. One of the reasons for this high interest rate is inflation. When economic calculations are made, it is acceptable to work in terms of either "nominal" dollars—i.e., dollars having the value of the year in which they are spent (or received)—or in "real" dollars—i.e., dollars having present-day value. Provided that the correct cost of capital is used in the analysis, the same total discounted cost is obtained whether nominal or real dollars are used. (This does assume that inflation proceeds at a constant annual rate.)

To illustrate the method for obtaining the cost of capital if all calculations are done in present-day—i.e., real—dollars, consider the following.

ASSUMING NO INFLATION

I put $100 in the bank today, leave it for 1 year, and the bank will pay me 5% interest for doing so. Thus, at the end of the year I can still buy $100 worth of goods plus an item costing $5.00. My return for doing without my $100 for 1 year is an item costing $5 (and I can still have goods costing $100 if I wish).

ASSUME INFLATION AT 10% P.A.

I put $100 in the bank today. To obtain a "real" return of 5% by foregoing the use of this $100 today requires that I can still buy $100 worth of goods—which in 1 year would cost $110 since inflation occurs at 10%, plus the item originally costing $5.00, which 1 year later would cost $5.00 + 10% of $5.00, which is $5.50. Thus, at the end of the year, I need to have $100 + $10 + $5 + $0.50 = $115.50. The interest required on my $100 investment, then, is

$$\$10 + \$5 + \$0.50$$
$$= \theta + i + i\theta$$
$$= 10\% + 5\% + 0.5\%$$
$$= 15.5\%$$

where θ = inflation rate and i = interest rate. Alternatively, if the interest rate today for discounting is 15.5% and inflation is at 10% then the "real" interest rate is:

$$i = \left(\frac{1 + 0.155}{1 + 0.10}\right) - 1$$

$$= \left(\frac{1.155}{1.10}\right) - 1$$

$$= 1.05 - 1$$

$$= 0.05 \text{ (or 5\%)}$$

Formally, the appropriate cost of capital when working in present-day dollars is:

$$i = \frac{1 + \text{Cost of capital (as a decimal and with inflation)}}{1 + \text{Inflation rate (as a decimal)}}$$

AN EXAMPLE TO ILLUSTRATE THE USE OF REAL AND NOMINAL DOLLARS

Assume we are in the year 2020:

Cost of truck in 2000 = $75,000

Maintenance cost for a new truck in 2020
$$= \$5000$$

Maintenance cost for a 1 year old truck in 2020
$$= \$10,000$$

Maintenance cost for a 2 year old truck in 2020
$$= \$15,000$$

Assuming the real cost of capital = 15%, then the total discounted cost of the above series of cash flow is:

$$\$75,000 + \$5000 + \$10,000\left(\frac{1}{1+.15}\right)^1$$

$$+ \$15,000\left(\frac{1}{1+.15}\right)^2$$

$$= \$75,000 + \$5000 + \$10,000\,(0.8696)$$
$$+ \$15,000\,(0.8696)^2$$
$$= \$75,000 + \$5000 + \$8696 + \$11,342$$
$$= \$100,038$$

Assuming inflation now occurs at an average rate of 10% per annum, then the cost of capital would be:

$$0.15 + 0.10 + 0.10(0.15) = 0.265$$

The cost data would now be:

Cost of truck in 2020 = $75,000

Maintenance cost of a new truck for 1 year in 2020
$$= \$5000$$

Maintenance cost of a 1-year-old truck for 1 year in 2021
$$= \$10,000 \ (1 + 0.01) = \$11,000$$

Maintenance cost of a 2-year-old truck for 1 year in 2022
$$= \$15,000 \ (1 + 0.1)^2 = \$18,150$$

Using a cost of capital of 26.5%, the total discounted cost would now be:

$$\$75,000 + \$5000 + \$11,000 \left(\frac{1}{1 + 0.265} \right)^1$$

$$+ \ \$18,150 \left(\frac{1}{1 + 0.265} \right)^2$$

$$= \$75,000 + \$5000 + \$11,000 \ (0.7905)$$
$$+ \ \$18,150 \ (0.6249)$$
$$= \$75,000 + \$5000 + \$8696 + \$11,343$$
$$= \$100,038$$

(which is identical to the result obtained when present-day dollars were used along with an "uninflationary" cost of capital.

EQUIVALENT ANNUAL COST

The economic-life model gives a dollar cost that is a consequence of an infinite chain of replacements, or by modification (see following) for the first N cycles. To ease interpretation of this total discounted cost, it is useful to convert the total discounted costs associated with the economic life to an Equivalent Annual Cost (EAC).

In the calculations performed above, the total discounted cost over the 3 years was $100,038 when the cost of capital was 15%. Using a quantity called the "capital recovery factor" (CRF) it is possible to convert the $100,038 to an equivalent annual cost. The formula for calculating the CRF is:

$$CRF = \frac{i(1 + i)^n}{(1 + i)^n - 1}$$

Rather than use the above formula, most books dealing with financial analysis include tables of CRFs for a variety of interest rates. In such tables one would find for $i = 15\%$ and $n = 3$ years the CRF = 0.4380 and the EAC is $100,038 (0.4380) = $43,816.644 p.a. This means that a constant payment of $43,816.644 p.a. would result in a total discounted cost (TDC) of $100,038 if the cost of capital is taken at 15%.

Check:

$$TDC = \$43,816.644\left(\frac{1}{1 + 0.15}\right)^1$$

$$+ \$43,816.644\left(\frac{1}{1 + 0.15}\right)^2$$

$$+ \$43,816.644\left(\frac{1}{1 + 0.15}\right)^3$$

$$= \$38,101.43 + \$33,131.678 + \$28,810.152$$

$$= \$100,043$$

(given rounding errors, this is equal to $100,038).

Appendix 10: Optimal Number of Workshop Machines to Meet a Fluctuating Workload

STATEMENT OF PROBLEM

From time to time, jobs requiring the use of workshop machines (say, lathes) are sent from various production facilities within an organization to the maintenance workshop. Depending on the workload of the workshop, these jobs will be returned to production after some time has elapsed. The problem is to determine the optimal number of machines that minimizes the total cost of the system. This cost has two components: the cost of the workshop facilities and the cost of downtime incurred by jobs waiting in the workshop queue and then being repaired.

CONSTRUCTION OF MODEL

1. The arrival rate of jobs to the workshop requiring work on a lathe is Poisson-distributed with arrival rate λ.
2. The service time a job requires on a lathe is negative-exponentially distributed with mean $1/\mu$.
3. The downtime cost per unit time for a job waiting in the system (i.e., being served or in the queue) is C_d.
4. The cost of operation per unit time for one lathe (either operating or idle) is C_l.
5. The objective is to determine the optimal number of lathes n to minimize the total cost per unit time $C(n)$ of the system.

$C(n)$ = Cost per unit time of the lathes

      + Downtime cost per unit time due to

                    jobs being in the system

Cost per unit time of the lathes

= Number of lathes

  $\times$ Cost per unit time per lathe

= nC_1

Downtime cost per unit time of jobs

        being in the system

= Average wait in the system per job

  $\times$ Arrival rate of jobs in the system per unit time

  $\times$ Downtime cost per unit time per job

= $W_s \lambda C_d$

where W_s = mean wait of a job in the system. Hence

    $C(n) = nC_l + W_s \lambda C_d$

This is a model of the problem relating number of machines n to total cost $C(n)$.

Appendix 11: Optimal Size of a Maintenance Workforce to Meet a Fluctuating Workload, Taking into Account Subcontracting Opportunities

STATEMENT OF PROBLEM

The workload for the maintenance crew is specified at the beginning of a period—say, a week. By the end of the week the entire workload must be completed. The size of the workforce is fixed—there is a fixed number of manhours available per week. If demand at the beginning of the week requires fewer manhours than the fixed capacity, no subcontracting takes place. If, however, the demand is greater than the capacity, then the excess workload is subcontracted and returned from the subcontractor by the end of the week. Two sorts of costs are incurred:

1. Fixed cost depending on the size of the workforce.
2. Variable cost depending on the mix of internal/external workload

As the fixed cost is increased through increasing the size of the workforce, there is less chance of subcontracting being necessary. However, there may often be occasions when fixed costs will be incurred yet demand low, i.e., considerable underutilization of workforce. The problem is to determine the optimal size of the workforce to meet a fluctuating demand in order to minimize expected total cost per unit time.

CONSTRUCTION OF MODEL

1. The demand per unit time is distributed according to a probability density function $f(r)$, where r is the number of jobs.
2. The average number of jobs processed per worker per unit time is m.
3. The total capacity of the workforce per unit time is nm, where n is the number of staff in the workforce.
4. The average cost of processing one job by the workforce is C_1.
5. The average cost of processing one job by the subcontractor is C_s.
6. The fixed cost per man per unit time is C_f.

The basic conflicts of this problem are illustrated in Figure A.11.1, in which it is seen that the expected total cost per unit time $C(n)$ is

$C(n)$ = Fixed cost per unit time

    + Variable internal processing per unit time

    + Variable subcontracting processing

       cost per unit time

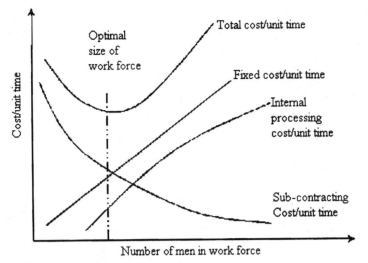

Figure A.11.1

Fixed cost per unit time = Size of workforce
$$\times \text{ Fixed cost per person}$$
$$= nC_f$$

Variable internal processing cost per unit time =
Average number of jobs processed
internally per unit time
$$\times \text{ Cost per job}$$

Now, number of jobs processed internally per unit time will be

1. Equal to the capacity when demand is greater than capacity
2. Equal to demand when demand is less than, or equal to, capacity

Probability of no. 1 $= \displaystyle\int_{nm}^{\infty} f(r)\,dr$

Probability of no. 2 $= \displaystyle\int_{-\infty}^{nm} f(r)\,dr = 1 - \int_{nm}^{\infty} f(r)\,dr$

When no. 2 occurs, the average demand will be

$$\int_0^{nm} f(r)dr \left| \int_0^\infty f(r)dr \right.$$

Therefore, variable internal processing cost per unit time is

$$\left(nm \int_{nm}^\infty f(r)dr + \frac{\int_0^{nm} rf(r)dr}{\int_0^{nm} rf(r)dr} \int_0^{nm} rf(r)dr \right) C_w$$

Variable subcontracting processing cost per unit time =
Average number of jobs processed
externally per unit time
$\times$ Cost per job

Now, the number of jobs processed externally will be

1. Zero when the demand is less than the workforce capacity
2. Equal to the difference between demand and capacity when demand is greater than capacity.

Probability of no. 1 $= \int_0^{nm} f(r)dr$

Probability of no. 2 $= \int_{nm}^\infty f(r)dr = 1 - \int_0^{nm} f(r)dr$

When no. 2 occurs, the average number of jobs subcontracted is:

$$\int_{nm}^\infty (r - nm)f(r)dr \left| \int_{nm}^\infty f(r)dr \right.$$

Therefore, variable subcontracting processing cost per unit time is

$$\left(0 \times \int_0^{nm} f(r)dr + \frac{\int_{nm}^{\infty} (r - nm)f(r)dr}{\int_{nm}^{\infty} f(r)dr} \int_{nm}^{\infty} f(r)dr \right) C_s$$

Therefore

$$c(n) = nC_f + \left(nm \int_{rm}^{\infty} f(r)dr + \int_0^{nm} f(r)dr \right) C_w$$

$$+ \int_{nm}^{\infty} (r - nm)f(r)dr$$

This is a model of the problem relating workforce size n to total cost per unit time $C(n)$.

Appendix 12: Fuzzy Logic

Fuzzy logic terminology: degree of membership, universe of discourse, membership function/fuzzy subset. In Figure A.12.1, as an example, the universe of discourse is the set of people. To each person in the universe of discourse, we assign a degree

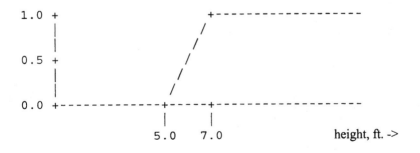

Figure A.12.1

of membership in the fuzzy subset TALL. George, who is 4′ 8″ (by substitution in the membership function above), has degree of membership in TALL of 0. Sam, 5′ 9″, has a degree of tallness of 0.38. Kareem, at 7′ 2″, is tall: dom. 1.0.

Appendix 13: Neural Nets

Here is the neural net example from Chapter 10 (which illustrates many of the concepts) taken from Donald R. Tveter's manual to the professional version of his Backprop Algorithm software package (Ref. 4 in Chapter 10). A three-layer network to provide XOR functionality after suitable "training" is described below (how it gets trained is explained later).

A THREE-LAYER MULTILAYER PERCEPTRON

Figure A.13.1 is a 2-1-1 network (multilayer perceptron, or MLP) with extra input-output connections (better for some problems like XOR). To get an answer for the case $x = 1$ and

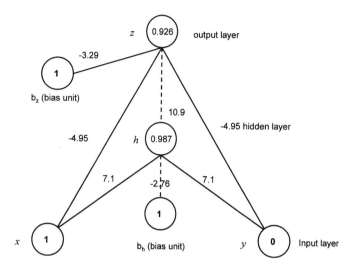

Figure A.13.1

$y = 0$, first compute the sum of inputs to the hidden layer neuron h: $s = (1 \times 7.1) + (1 \times -2.76) + (0 \times 7.1) = 4.34$. Then apply the activation function $v = 1/(1 + e^{-S})$ to compute the activation value, $v = 1/(1 + e^{-4.34}) = 0.98\%$. The output neuron is computed similarly: $s = (1 \times -4.95) + (0 \times -4.95) + (0.98\% \times 10.9) + (1 \times -3.29) = 2.52$ and $v = 0.926$. Of course, 0.926 is not quite 1, but for this example it's close enough. There are other activation functions you can use that make it easier to get closer to the targets. Here is the

$$\text{net}_j = \sum_{i=1,n} w_{ij}o_i \qquad o_j = f(\text{net}_j)$$

where net_j = net input to neuron j; f = the activation function; w_{ij} = the weight between neurons j and i. Now let's see how the network gets trained. The method is back-propagation. These are the steps:

1. Start with arbitrary values for w_{ij}. In this case the network weights all start out at 0 (a bad idea in general, but it works here). The activation values for the

hidden and output neurons are both 0.5 as calculated by the above two formulas.

2. Put one of the patterns to be learned on the input units.
3. Find the values of the hidden and output units.
4. Find out how large the error is on the output unit.
5. Use one of the back-propagation formulas to adjust the weights leading into the output unit.
6. Use another formula to find the errors for the hidden layer unit.
7. Adjust the weights leading into the hidden layer unit via another formula.
8. Repeat steps 2 to 7 for the second, third, and fourth XOR patterns.

Even after all these changes, the answers will be only a little closer to the right answers, and the whole process must be repeated many times. Each time all the patterns in the problem are used is an *epoch*. Here are the formulas:

$$\delta_k = (t_k - o_k)f'(\text{net}_k)$$

$$f = \frac{1}{1 + e^{-\text{net}_k}}$$

$$f' = o_k(1 - o_k)$$

$$w_{jk} \leftarrow w_{jk} + \eta\delta_k o_j$$

where: δ_k = error signal; t_k = target for unit k; f' = derivative of activation function f, η = learning rate = 0.1. Here is the first set of calculations:

$$\delta_z = (1 - 0.5) * 0.5 * (1 - 0.5) = 0.125$$

$$w_{xz} \leftarrow 0 + 0.1 * 0.125 * 1 = 0.0125$$

$$w_{yz} \leftarrow 0 + 0.1 * 0.125 * 0 = 0$$

$$w_{hz} \leftarrow 0 + 0.1 * 0.125 * 0.5 = 0.00625$$

$$w_{bz} \leftarrow 0 + 0.1 * 0.125 * 1 = 0.0125$$

$$\delta_j = f'(\text{net}_j) \sum_k \delta_k w_{kj}$$

where δ_j = error for a hidden unit j. Therefore, $\delta_h = o_h(1 - o_h)\delta_{zh}$ = 0.5 * (1 − 0.5) * 0.125 * 0.00625 = 0.000195313.

$w_{ij} \leftarrow w_{ij} + \eta\delta_j o_i$ is weight-change formula between hidden unit j and input unit i. Therefore, w_{hx} = 0 + 0.1 * 0.000195313 * 1 = 0.0000195313; w_{hbh} = 0 + 0.1 * 0.000195313 * 1 = 0.0000195313. The activation value for the output layer will now calculate to 0.507031, a long way from our target. If we do the same for the other three patterns we get:

x	y	z_desired	z_actual
1	0	1	0.499830
0	0	0	0.499893
0	1	1	0.499830
1	1	0	0.499768

The eight-step backprop training algorithm as given above, although it works, is, strictly speaking, considered wrong by mathematicians. Rather than changing the weights leading into the output layer before propagating the error back, it is mathematically more rigorous to compute the weight changes for the upper layer of weights and then, without changing them, compute weight changes for the lower layer of weights. Then, with all the changes in hand, make the changes to both sets.

Sadly, to get the outputs to within 0.1 requires 20,862 iterations (epochs)—a very long time, especially for such a short problem. Fortunately, a lot of things can be done to speed up the training so that the number of epochs can be reduced to around 12–20. The simplest would be to increase the learning rate η (Figure A.13.2).

η	iterations
0.1	20,862
0.5	2,455
1.0	1,060
2.0	480
3.0	(fails)

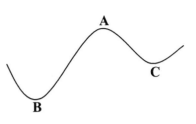

Figure A.13.2

FAILURE OF A NEURAL NET TO TRAIN

An unfortunate problem with back-propagation is that, when the learning rate is too large, the training can fail the figure above. In this case, with $\eta = 3$ after 10,000 epochs (iterations), $x = 1$, $y = 1$ resulted in $z = 1$, which is not an exclusive *or*. The geometric interpretation (see Figure A.13.2) is that when the network tries to make the error go down, it may get stuck in a valley that is not the lowest possible valley. That is a local minimum, which is not the global minimum that would give perfect results. Once you go down into a shallow valley, if there is no path downhill that will get you to the deeper valley you are permanently stuck with a bad answer. There are ways to break out of a local minimum, but in real-world problems you never know when you've found the best (global) minimum. You only hope to get a very deep minimum.

Appendix 14: Optimal Interval Between Preventive Replacements of Equipment Subject to Breakdown

STATEMENT OF PROBLEM

Equipment is subject to sudden failure, and when failure occurs the equipment must be replaced. Since failure is unexpected, it is not unreasonable to assume that a failure replacement is more costly than a preventive replacement. For example, when a preventive replacement is planned, arrangements are made to perform it without unnecessary delays; a failure may cause damage to other equipment. To reduce the number of failures, preventive replacements can be scheduled to occur at specified intervals. However, a balance is required between the amount spent on the preventive replacements and their resulting benefits, i.e., reduced failure replacements.

In this appendix it will be assumed, not unreasonably, that we are dealing with a long period of time over which the equipment is to be operated and relatively short intervals between the preventive replacements. When this is the case, we need consider only one cycle of operations and develop a model for the cycle. If the interval between the preventive replacements is long, it is necessary to use the discounting approach and the series of cycles must be included in the model.

The replacement policy calls for preventive replacements to occur at fixed intervals of time and failure replacements to occur when necessary, and we want to determine the optimal interval between the preventive replacements to minimize the total expected cost of replacing the equipment per unit time.

CONSTRUCTION MODEL

1. C_p is the cost of a preventive replacement.
2. C_j is the cost of a failure replacement.
3. $f(t)$ is the probability density function of the equipment's failure times.
4. The replacement policy is to perform preventive replacements at constant intervals of length t_p, irrespective of the age of the equipment, and failure replacements occur as many times as required in interval $(0, t_p)$. The policy is illustrated in Figure A.14.1.

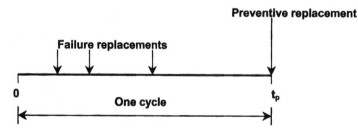

Figure A.14.1

5. The objective is to determine the optimal interval between preventive replacements to minimize the total expected replacement cost per unit time.

The total expected cost per unit time for preventive replacement at time t_p, denoted $C(t_p)$, is

$$C(t_p) = \frac{\text{Total expected cost in interval } (0,\ t_p)}{\text{Length of interval}}$$

Total expected cost in interval $(0,\ t_p)$ = Cost of a preventive replacement + Expected cost of failure replacements = $C_p + C_f H(t_p)$ where $H(t_p)$ is the expected number of failures in interval $(0,\ t_p)$.

Length of interval = t_p

Therefore

$$C(t_p) = \frac{C_p + C_f H(t_p)}{t_p}$$

This is a model of the problem relating replacement interval t_p to total cost $C(t_p)$.

Appendix 15: Optimal Preventive Replacement Age of Equipment Subject to Breakdown

STATEMENT OF PROBLEM

This problem is similar to that of Appendix 14 except that instead of making preventive replacements at fixed intervals, thus incurring the possibility of performing a preventive replacement shortly after a failure replacement, the time at which the preventive replacement occurs depends on the age of the equipment. When failure occurs, failure replacements are made.

Again, the problem is to balance the cost of the preventive replacements against their benefits. We do this by determining the optimal preventive replacement age for the equipment to minimize the total expected cost of replacements per unit time.

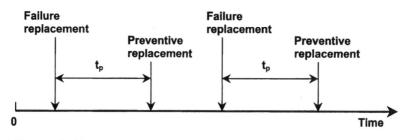

Figure A.15.1

CONSTRUCTION OF MODEL

1. C_p is the cost of a preventive replacement.
2. C_f is the cost of a failure replacement.
3. $f(t)$ is the probability density function of the failure times of the equipment.
4. The replacement policy is to perform preventive replacements once the equipment has reached a specified age at t_p, plus failure replacements when necessary. The policy is illustrated in Figure A.15.1.
5. The objective is to determine the optimal replacement age of the equipment to minimize the total expected replacement cost per unit time.

In this problem, there are two possible cycles of operation: one cycle is determined by the equipment's reaching its planned replacement age t_p, the other by the equipment's ceasing to operate due to a failure occurring before the planned replacement time. These two possible cycles are illustrated in Figure A.15.2.

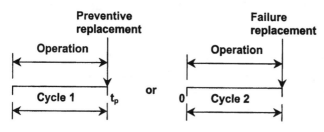

Figure A.15.2

The total expected replacement cost per unit time $C(t_p)$ is

$$C(t_p) = \frac{\text{Total expected replacement cost per cycle}}{\text{Expected cycle length}}$$

Total expected replacement cost per cycle

= Cost of a preventive cycle

× Probability of a preventive cycle

+ Cost of failure cycle

× Probability of a failure cycle

= $C_p R(t_p) + C_f[1 - R(t_p)]$

Remember: if $f(t)$ is as illustrated in Figure A.15.3, then the probability of a preventive cycle equals the probability of failure occurring after time t_p; that is, it is equivalent to the shaded area, which is denoted $R(t_p)$.

The probability of a failure cycle is the probability of a failure occurring before time t_p, which is the unshaded area area of Figure A.15.3. Since the area under the curve equals unity, then the unshaded area is $[1 - R(t_p)]$.

Expected cycle length = Length of a preventive cycle

× Probability of a preventive cycle

+ Expected length of a failure cycle

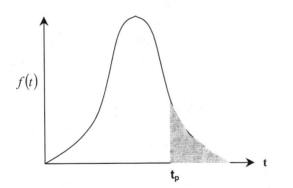

Figure A.15.3

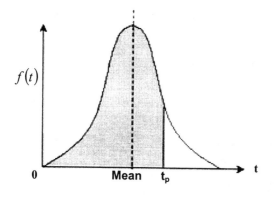

Figure A.15.4

$\times$ Probability of a failure cycle

$= t_p \times R(t_p)$

$+$ (Expected length of a failure cycle) $\times$ $[1 - R(t_p)]$

To determine the expected length of a failure cycle, consider Figure A.15.4. The mean time to failure of the complete distribution is

$$\int_{-\infty}^{\infty} tf(t)dt$$

which for the normal distribution equals the mode (peak) of the distribution. If a preventive replacement occurs at time t_p, then the mean time to failure is the mean of the shaded portion of Figure A.15.4 since the unshaded area is an impossible region for failures. The mean of the shaded area is

$$\int_{-\infty}^{t_p} \frac{tf(t)dt}{1 - R(t_p)}$$

denoted $M(t_p)$. Therefore, the expected cycle length $= t_p \times R(t_p) + M(t_p) \times [1 - R(t_p)]$

$$C(t_p) = \frac{C_p \times R(t_p) + C_f \times [1 - R(t_p)]}{t_p \times R(t_p) + M(t_p) \times [1 - R(t_p)]}$$

This is now a model of the problem relating replacement age t_p to total expected replacement cost per unit time.

Appendix 16: Optimal Preventive Replacement Age of Equipment Subject to Breakdown, Taking into Account Time Required to Effect Failure and Preventive Replacements

STATEMENT OF PROBLEM

The problem definition is identical to that of Appendix 15 except that, instead of assuming that the failure and preventive replacements are made instantaneously, the time required to make these replacements is taken into account.

The optimal preventive replacement age of the equipment is again taken as the age that minimizes the total expected cost of replacements per unit time.

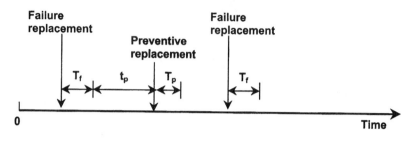

Figure A.16.1

CONSTRUCTION OF MODEL

1. C_p is the cost of a preventive replacement.
2. C_f is the cost of a failure replacement.
3. T_p is the time required to make a preventive replacement.
4. T_f is the time required to make a failure replacement.
5. $f(t)$ is the probability density function of the failure times of the equipment.
6. $M(t_p)$ is the mean time to failure when preventive replacement occurs at time t_p.
7. The replacement policy is to perform a preventive replacement once the equipment has reached a specified age at t_p, plus failure replacements when necessary. The policy is illustrated in Figure A.16.1.
8. The objective is to determine the optimal preventive replacement age of the equipment to minimize the total expected replacement cost per unit time.

As was the case for the problem in Appendix 15, there are two possible cycles of operation, which are illustrated in Figure A.16.2.

The total expected replacement cost per unit time, denoted $C(t_p)$, is:

$$C(t_p) = \frac{\text{Total expected replacement cost per cycle}}{\text{Expected cycle length}}$$

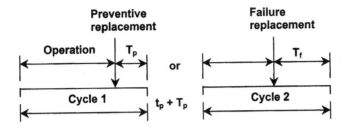

Figure A.16.2

Total expected replacement cost per cycle:

$$= C_p \times R(t_p) + C_f[1 - R(t_p)]$$

Expected cycle length:

$$= \text{Length of a preventive cycle}$$
$$\times \text{ Probability of a preventive cycle}$$
$$+ \text{ Expected length of a failure cycle}$$
$$\times \text{ Probability of a failure cycle}$$
$$= (t_p + T_p)R(t_p) + [M(t_p) + T_f][1 - R(t_p)]$$

$$C(t_p) = \frac{C_p R(t_p) + C_f[1 - R(t_p)]}{(t_p + T_p)R(t_p) + [M(t_p) + T_f][1 - R(t_p)]}$$

This is a model of the problem relating preventive replacement age t_p to the total expected replacement cost per unit time.

Appendix 17: Optimal Replacement Interval for Capital Equipment: Minimization of Total Cost

STATEMENT OF PROBLEM

This problem is introduced in Section 11.3. The objective is to determine the replacement interval that minimizes the total cost of maintenance and replacement over a long period where the trend in costs is taken to be discrete, rather than continuous.

CONSTRUCTION OF MODEL

1. A is the acquisition cost of the capital equipment.
2. C_i is the cost of maintenance in the ith period from

new, assumed to be paid at the end of the period $i =$
1, 2, . . . , n.
3. S_i is the resale value of the equipment at the end of
   the ith period of operation $i = 1, 2, \ldots, n$.
4. r is the discount rate.
5. n is the age in periods of the equipment when replaced.
6. $C(n)$ is the total discounted cost of maintaining and
   replacing the equipment (with identical equipment)
   over a long period of time with replacements occurring at intervals of n periods.
7. The objective is to determine the optimal interval between replacements to minimize total discounted
   costs $C(n)$.

The replacement policy is illustrated in Figure A.17.1.
Consider first the cycle of operation. The total discounted
cost over the first cycle of operation, with equipment already
installed, is

$$C_1(n) = C_1 r + C_2 r^2 + C_3 r^3 + \cdots + C_n r^n + A r^n - S_n r^n$$

$$= \sum_{i=1}^{n} C_i r^i + r^n (A - S_n)$$

For the second cycle, the total cost discounted to the start of
the second cycle is

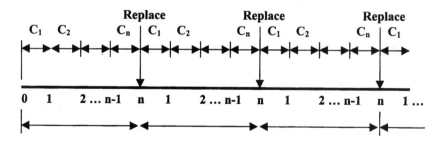

Figure A.17.1

$$C_2(n) = \sum_{i=1}^{n} C_i r^i + r^n(A - S_n)$$

Similarly, the total costs of the third and fourth cycle, and so on, discounted back to the start of their cycle, can be obtained.

The total discounted costs, when discounting is taken to the start of the operation, i.e., at time 0, is

$$C(n) = C_1(n) + C_2(n)r^n + C_3(n)r^{2n} + \cdots + C_n(n)r^{(n-1)n} + \cdots$$

Since $C_1(n) = C_2(n) = C_3(n) = \cdots = C_n(n) = \ldots$, we have a geometric progression that gives, over an infinite period:

$$C(n) = \frac{C_1(n)}{1 - r^n} = \frac{\displaystyle\sum_{i=1}^{n} C_i r^i + r^n(A - S_n)}{1 - r^n}$$

This is a model of the problem relating replacement interval n to total costs.

Appendix 18: The Economic-Life Model Used in PERDEC

The economic life model used in PERDEC is:

$$\text{EAC}(n) = \left[\frac{A + \sum_{i=0}^{n} C_i r^i - S_n r^n}{1 - r^n} \right] * i$$

where
A = acquisition cost
C_i = O&M costs of equipment in its ith year of life, assuming payable at the start of year, $i = 1, 2, \ldots, n$
r = Discount factor
S_n = Resale value of equipment of age n years
n = Replacement age
$C(n)$ = Total discounted cost for a chain of replacements every n years

Appendix 19: Economic Life of Passenger Buses

The purpose of this appendix is to demonstrate how the standard economic life model for equipment replacement can be modified slightly to enable the economic life of buses to be determined, taking into account declining utilization of a bus over its life. Specifically, new buses are highly utilized to meet base load demand while older buses are used to meet peak demands, such as during rush hour.

The case study describes a fleet of 2000 buses whose annual fleet demand is 80 million kilometers. The recommendations resulting from the study were implemented and substantial savings were reported.

INTRODUCTION

The study took place in Montreal, Canada, where Montreal's Transit Commission is responsible for providing bus services to approximately 2 million people. To meet the bus schedules requires a fleet of 2000 buses undertaking approximately 80,000,000 kilometers/year. Montreal has the third largest fleet of North American bus operators, behind New York City and Chicago. The bulk of the buses used in Montreal were standard 42-seat GMC buses.

The objective of the study was to analyze bus operations and maintenance costs to determine the economic life of a bus and to identify a steady-state replacement policy, i.e., one in which a fixed proportion of the fleet would be replaced on an annual basis.

THE ECONOMIC-LIFE MODEL

Figure A.19.1 illustrates the standard conflicts associated with capital-equipment replacement problems.

The standard economic life model is

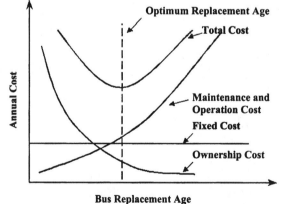

Figure A.19.1

$$C(n) = \frac{A + C_1 + \sum_{i=2}^{n} C_i r^{i-1} - S_n r^n}{1 - r^n}$$

where

A = Acquisition cost

C_1 = Operations and maintenance (O&M) cost of a bus in its first year of life

C_i = O&M cost of a bus in its ith year of life, assuming payable at the start of the year, $i = 2, 3, \ldots, n$

r = Discount factor

S_n = Resale value of a bus of age n years

n = Replacement age

$C(n)$ = Total discounted cost for a chain of replacements every n years

The total discounted cost can be converted to an Equivalent Annual Cost (EAC) by the Capital Recovery Factor, which in this case is i, the interest rate appropriate for discounting. Note:

$$r = \frac{1}{1 + i}$$

DATA ACQUISITION

Acquisition cost: In terms of 1980 dollars, the acquisition cost of a bus was $96,330.

Resale value: The policy being implemented by Montreal Transit was replacement of a bus on a 20-year-old cycle, at which age the value of the bus was $1000. Because there was considerable uncertainty in the resale value of a bus, for purposes of the study two extremes were evaluated: a high and low trend in resale value (Tables A.19.1 and A.19.2).

Table A.19.1   High
Trend in Resale Value

Replacement age (years)	Resale value ($)
1	77,000
2	65,000
3	59,000
4	54,000
5	50,000
6	46,000
7	42,000
8	38,000
9	34,000
10	31,000
11	28,000
12	25,000
13	22,000
14	19,000
15	16,000
16	13,000
17	10,000
18	7,000
19	4,000
20	1,000

INTEREST RATE

The interest rate appropriate for discount was uncertain. In the study a range (0% to 20%) was used to check the sensitivity of the economic life to variations in interest rate. The major conclusions of the study were based on an "inflation fee" interest rate of 6%.

OPERATIONS AND MAINTENANCE COST

O&M costs are influenced by both age of a bus and its cumulative utilization. O&M data were obtained for six cost catego-

Table A.19.2   High
Trend in Resale Value

Replacement age (years)	Resale value ($)
1	2000
2	2000
3	2000
4	2000
5	2000
6	2000
7	2000
8	2000
9	2000
10	2000
11	2000
12	2000
13	2000
14	2000
15	2000
16	2000
17	2000
18	2000
19	2000
20	1000

ries: fuel, lubrication, tires, oil, parts, and labor. Analysis of the costs identified the following trend:

$$c(k) = 0.302 + 0.723 \left(\frac{k}{10^6}\right)^2$$

where

k = Cumulative kilometers traveled by the bus since new

$c(k)$ = Trend in O&M cost in $/kilometer for a bus of age k kilometers

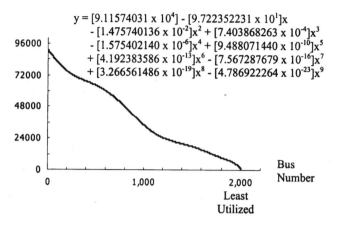

$$y = [9.11574031 \times 10^4] - [9.722352231 \times 10^1]x$$
$$- [1.475740136 \times 10^{-2}]x^2 + [7.403868263 \times 10^{-4}]x^3$$
$$- [1.575402140 \times 10^{-6}]x^4 + [9.488071440 \times 10^{-10}]x^5$$
$$+ [4.192383586 \times 10^{-13}]x^6 - [7.567287679 \times 10^{-16}]x^7$$
$$+ [3.266561486 \times 10^{-19}]x^8 - [4.786922264 \times 10^{-23}]x^9$$

Figure A.19.2

BUS UTILIZATION

Figure A.19.2 shows the trend line that was fitted to the relationship between bus utilization (km/yr) and bus age (newest to oldest). The reason for this relationship is that new buses are highly utilized to meet base load requirements, with the older buses being used to meet peak demands.

In the analysis that was undertaken, it was assumed that the relationship identified in Figure A.19.2 would be independent of the replacment age of the bus. For example, using the present policy of replacing buses on a 20-year cycle, in a steady state, 2000/20 = 100 buses would be replaced annually. The total work done by the newest 100 buses in their first year of life would then be:

$$\int_0^{100} Y\, dk = 8{,}636{,}059 \text{ kilometers}$$
$$= 86{,}361 \text{ km/bus}$$

SOLVING THE MODEL

If the economic replacement age of a bus is set at 20 years, then $C(20)$ can be evaluated using the previously cited data

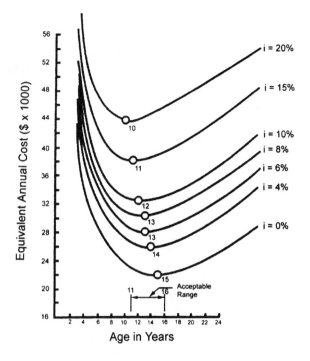

Figure A.19.3

and converting the annual utilization to annual cost by integrating the cost trend $c(k)$ over the appropriate utilization range. Thus for a bus in its first year of life the O&M cost would be:

$$\int_0^{86361} 0.302 + 0.723\left(\frac{k}{10^6}\right)^2 dk$$

Similarly, the O&M costs for the remaining 19 years can be calculated and inserted into the economic-life model $C(n)$ to enable $C(20)$ to be calculated. The whole process is then repeated for other possible replacement ages to identify the optimal n. Figure A.19.3 shows the results when i ranges from 0% to 20% and the low trend in resale value is used.

CONCLUSIONS

Montreal Transit decided to implement a 3-year bus purchase policy on the basis of a 16-year replacement age. The resultant savings approximated $4 million/year. The class of problem discussed here is typical of those found in many transport operations, for example:

- Haulage fleets undertaking both long-distance and local deliveries—new vehicles are used on long-haul routes; as they age, they are relegated to local delivery work.
- Stores with their own fleets of delivery vehicles, with peak demands around Christmas. The older vehicles in the fleet are retained to meet these predictable demands. Because of this unequal utilization, it is necessary to evaluate the economic life of the vehicle by viewing the fleet as a whole rather than focus on individual vehicles.

REFERENCES

1. B. W. Simms, B. G. Lamarre, A. K. S. Jardine, A. Boudreau. Optimal Buy, Operate and Sell Policies for Fleets of Vehicles. European Journal of Operational Research 1984; 15(2):183–195. (This paper discusses a dynamic programming/linear programming approach to the problem that enabled alternative utilization policies to be incorporated into the model.)
2. AGE/CON User's Manual. www.oliver-group.com (AGE/CON is a software package used for vehicle fleet replacement decisions. It includes the routines for handling problems of the class described in this appendix).

Appendix 20: Optimal Replacement Age of an Asset, Taking into Account Tax Considerations

The formula is

$$
\begin{aligned}
\mathrm{NPV}(k) = A &\left[1 - \frac{td}{i + d} \left(\frac{1 + \dfrac{i}{2}}{1 + i} \right) \right] \\
&- \frac{itA\left(1 - \dfrac{d}{2} \right)(1 - d)^{k-1}}{(i + d)(1 + i)^k} \\
&- \frac{S_k(1 - t)}{(1 + i)^k} \\
&+ \sum_{j=1}^{k} C_j \frac{(1 - t)}{(1 + i)^{j-1}}
\end{aligned}
$$

where

i = Interest rate
d = Capital Cost Allowance rate
t = Corporation tax rate
A = Acquisition cost
S_k = Resale price
C_j = O&M cost in the jth year
NPV(k) = Net Present Value in the kth year

$A(1 - d/2)(1 - d)^{k-1}$ is the nondepreciated capital cost.

The equivalent annual cost (EAC) is then obtained by multiplying the NPV(k) by the Capital Recovery Factor

$$\frac{i(1 + i)^k}{(1 + i)^k - 1}$$

Appendix 21: Optimal Replacement Policy for Capital Equipment, Taking into Account Technological Improvement—Infinite Planning Horizon

STATEMENT OF PROBLEM

For this replacement problem, it is assumed that, once the decision has been made to replace the current asset with the technologically improved equipment, this equipment will continue to be used and a replacement policy (periodic) will be required for it. It will also be assumed that replacement will continue to be made with the technologically improved equipment. Again we wish to determine the policy that minimizes total discounted costs of maintenance and replacement.

CONSTRUCTION OF MODEL

1. n is the economic life of the technologically improved equipment.
2. $C_{p,t}$ is the maintenance cost of the present equipment in the ith period from now, payable at time i: $i = 1$, $2, \ldots, n$.
3. $S_{p,t}$ is the resale value of the present equipment at the end of the ith period from now; $i = 1, 2, \ldots, n$.
4. A is the acquisition cost of the technologically improved equipment.
5. $C_{t,j}$ is the maintenance cost of the technologically improved equipment in the jth period after its installation and payable at time j; $j = 1, 2, \ldots, n$.
6. $S_{t,j}$ is the resale value of the technologically improved equipment at the end of its jth period of operation; $j = 0, 1, 2, \ldots, n$. ($j = 0$ is included so that we can then define $S_{t,o} = A$. This then enables Ar^n in the model to be canceled if no change is made). Note that it is assumed that, if a replacement is to be made at all, it is with the technologically improved equipment. This is not unreasonable since it may be that the equipment currently in use is no longer on the market.
7. r is the discount factor.
8. The replacement policy is illustrated in Figure A.21.1.

The total discounted cost over a long period of time with replacement of the present equipment at the end of T period of operation, followed by replacements of the technologically improved equipment at intervals of n, is

$$C(T, n) = \text{Costs over interval } (0, T) + \text{Future costs}$$

$$\text{Costs over interval } (0, T) = \sum_{i=1}^{n} C_{p,t} r^i - S_{p,T} r^T + Ar^T$$

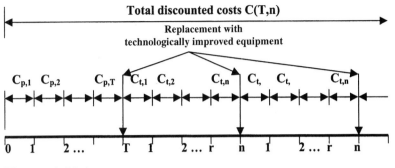

Figure A.21.1

Future costs, discounted to time T, can be obtained by the method described in Appendix 17, where the economic life of equipment is calculated. We replace C_i with $C_{t,j}$ to obtain

$$C(n) = \frac{\sum_{i=1}^{n} C_{t,j} r^i + r^n(A - S_n)}{1 - r^n}$$

Therefore, $C(n)$ discounted to time zero is $C(n)r^T$ and

$$C(T,n) = \sum_{i=1}^{T} C_{p,t} r^i - S_{p,T} r^T + A r^T + \left(\frac{\sum_{j=1}^{n} C_{t,j} r^i + r^n(A - S_n)}{1 - r^n} \right) r^T$$

This is a model of the problem relating change over time to technologically improved equipment T and economic life of new equipment n to total discounted costs $C(T, n)$.

Appendix 22: Optimal Replacement Policy for Capital Equipment, Taking into Account Technological Improvement—Finite Planning Horizon

STATEMENT OF PROBLEM

When determining a replacement policy, there may be on the market equipment that is, in some way, a technological improvement on the equipment currently being used. For example, O&M costs may be lower, throughput may be greater, and/or quality of output may be better. The problem discussed in this appendix is how to determine when, if at all, to take advantage of the technologically improved equipment.

It will be assumed that there is a fixed period of time from now during which equipment will be required and, if replacement is with the new equipment, then this equipment will remain in use until the end of the fixed period. The objective will be to determine when to make the replacement, if at all, to minimize total discounted costs of maintenance and replacment.

CONSTRUCTION OF MODEL

1. n is the number of operating periods during which equipment will be required.
2. The objective is to determine the value of T at which replacement should take place with the new equipment, $T = 0, 1, 2, \ldots, n$. The policy is illustrated in Figure A.22.1.

The total discounted cost over n periods, with replacement occurring at the end of the Tth period, is:

$C(T, n) =$ Discounted maintenance costs for present equipment over period $(0, T)$

$+$ Discounted maintenance costs for technologically improved equipment over period (T, n)

$+$ Discounted acquisition of new equipment

$-$ Discounted resale value of present

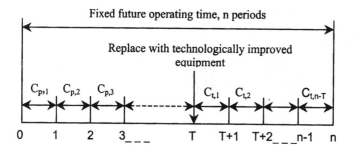

Figure A.22.1

equipment at end of the Tth period
— Discounted resale value of technologically
improved equipment at end of the nth period

$$= (C_{p,1}r^1 + C_{p,2}r^2 + C_{p,3}r^3 + \cdots + C_{p,T}r^T)$$
$$+ (C_{t,1}r^{T+1} + C_{t,2}r^{T+2} + \cdots + C_{t,n-T}r^n) + Ar^T$$
$$- (S_{p,T}r^T + S_{t,n-T}r^n)$$

Therefore

$$C(T, n) = \sum_{i=1}^{n} C_{p,t}r^i + \sum_{j=1}^{n-T} C_{t,j}r^{T+j} + Ar^T - (S_{p,T}r^T + S_{t,n-T}r^n)$$

This is a model of the problem relating replacement time T to total discounted cost $C(T)$.

Appendix 23: Optimal Inspection Frequency: Minimization of Downtime

STATEMENT OF PROBLEM

The problem in this section assumes that equipment breaks down from time to time and, to reduce the breakdowns, inspections and consequent minor modifications can be made. The problem is to determine the inspection policy that minimizes the total downtime per unit time incurred due to breakdowns and inspections.

CONSTRUCTION OF MODEL

1. Equipment failures occur according to the negative exponential distribution with mean time to failure

(MTTF) $= 1/\lambda$, where λ is the mean arrival rate of failures. (For example, if the MTTF $= 0.5$ years, then the mean number of failures per year $= 1/0.5 = 2$; i.e., $\lambda = 2$.)

2. Repair times are negative exponentially distributed with mean time $1/\mu$.

3. The inspection policy is to perform n inspections per unit time. Inspection times are negative exponentially distributed with mean time $1/i$.

The objective is to choose n to minimize total downtime per unit time.

The total downtime per unit will be a function of the inspection frequency n, denoted $D(n)$. Therefore:

$D(n) =$ Downtime incurred due to repairs per unit time

$+$ Down incurred due to inspection per unit time

$$= \frac{\lambda(n)}{\mu} + \frac{n}{i}$$

The equation above is a model of the problem relating inspection frequency n to total downtime $D(n)$.

Appendix 24: Maintenance Strategic Assessment (MSA) Questionnaire

This questionnaire should be filled out by the following personnel:

Department	Positions	How many?
Production	Managers	All
	Superintendents	All
	Supervisors	1 for each area (min)
	Operators	1 for each area (min)
Maintenance	Managers	All
	Superintendents	All
	Supervisors	1 for each area (min)
	Trades	1 for each trade (min)
Others	Managers	All

The results of this assessment will not be disclosed without the written permission of the firm being evaluated.

INSTRUCTIONS

Assign a score to each of the statements in the following questionnaire based on how well you think your maintenance organization adheres to the statement. The following rating scale must be used:

	Score
Strongly Agree	4
Mostly Agree	3
Partially Agree	2
Totally Agree	1
Do Not Understand	0

It is not necessary to add the scores up; that will be done when we enter them into our database. The final results will be presented on a scale of 100.

First, please tell us about yourself:

Name: _____   Job Title: _____

Plant/Site: _____   Division: _____

Primary responsibility	Department
_____ Management	_____ Maintenance
_____ Supervision	_____ Operations/Production
_____ Trades/Hourly	_____ Purchasing
_____ Administrative	_____ Tech/IT Support
_____ Other	_____ Other

1. MAINTENANCE STRATEGY

Statement	Score (4, 3, 2, 1, 0)
The maintenance department has a defined mission, mandate, and set of objectives that are well documented and understood by all personnel concerned.	
The maintenance mission statement and objectives clearly support a published statement of the company's objectives and goals, and the role of maintenance in achieving the company's objectives is understood.	
We have a long-term plan or strategy to guide maintenance improvement efforts which supports, and is linked to, the overall corporate strategy.	
We have a set of policies or guiding principles for maintenance. Maintenance is seen as a process not a function.	
Our approach to maintenance is proactive. We do our best to prevent breakdowns, and when something breaks we fix it immediately.	
Annual maintenance budget is prepared based on a long-term improvement plan, scheduled overhaul strategy, and history of equipment performance. Maintenance budget is related to expected performance and indications are provided as to the likely outcome if work is to be deferred.	
The maintenance budget has an allowance for any project work being done by the maintenance department. If not, project work is budgeted separately and accounted for outside of maintenance.	
Total (max. 28)	

Strongly Agree (4), Mostly Agree (3), Partially Agree (2), Totally Disagree (1), Do Not Understand (0)

2. ORGANIZATION/HUMAN RESOURCES

Statement	Score 4, 3, 2, 1, 0
Maintenance staffing level is adequate, highly capable, and experienced.	
Functions covering plant needs are fully defined, and our employees understand what is/is not expected of them, and organizational charts are current.	
The maintenance organization is mostly decentralized and organized by area or product line.	
First-line supervisors are responsible for at least 12 to 15 maintenance workers.	
Adequate support staffs are available to allow supervisors to spend more than 75% of their time in direct support of their people.	
Overtime represents less than 5% of the total annual maintenance manhours. Overtime is not concentrated in one trade group or area, but it is well distributed.	
Regular technical training is provided to all employees and is more than 5 days/year/employee. Maintenance supervisors have also received formal supervisory training.	
A formal established apprenticeship program is employed to address the maintenance department's needs for qualified trades. Clear standards are set for completing the apprenticeship programs.	
Part of the pay is based on demonstrated skills and knowledge and/or results and productivity.	
Contractors are used to augment plant staff during shutdowns and/or for specific projects or specialized jobs. Their cost/benefit is periodically reviewed.	

Total (max. 40)

Strongly Agree (4), Mostly Agree (3), Partially Agree (2), Totally Disagree (1), Do Not Understand (0)

3. EMPLOYEE EMPOWERMENT

Statement	Score (4, 3, 2, 1, 0)
We don't have a "Command and Control" organization with highly disciplined procedures.	
Multiskilled tradespeople (e.g., electricians doing minor mechanical work, mechanics doing minor electrical work, etc.) is a key feature of the organization.	
Operators understand the equipment they run, perform minor maintenance activities like cleaning, lubricating, minor adjustments, inspections, and minor repairs (not generally requiring the use of tools).	
Supervisors regularly discuss performance and costs with their work teams.	
Continuous-improvement teams are in place and active.	
Much of the work is performed by self-directed work teams of operators, maintainers, and engineers.	
Maintenance is a part of the team involved during design and commission of equipment modifications or capital additions to the plant.	
Trades usually respond to callouts after hours. Operations can get needed support from maintenance trades quickly and with a minimum of effort.	
Callouts are performed by an on-shift maintainer who decides what support is needed without reference to a supervisor for guidance. Operations does not decide who will be called.	
Partnerships have been established with key suppliers and contractors; risk-sharing is a feature of these arrangements.	
Total (max. 40)	

Strongly Agree (4), Mostly Agree (3), Partially Agree (2), Totally Disagree (1), Do Not Understand (0)

4. MAINTENANCE TACTICS

Statement	Score 4, 3, 2, 1, 0
Less than 5% of the total maintenance work man-hours is devoted to emergenies (e.g., unscheduled shutdowns).	
Condition-based maintenance is favored over time- or cycle-based maintenance.	
Use of condition-based maintenance techniques such as vibration analysis, oil sampling, nondestructive testing (NDT), and performance monitoring is widespread.	
Preventive and/or predictive maintenance represents 60% or more of the total maintenance man-hours.	
Compliance with the PM program is high: 95% or more of the PM work is completed as scheduled.	
Results from PM inspections and failure history data are used to continually refine and improve effectiveness of the PM program.	
For new equipment we review the manufacturer's maintenance recommendations and revise them as appropriate for our specific operating environment and demands.	
We used a formal reliability-based program for determining the correct PM routines to perform. That program is still used for continuously fine-tuning and improving our PM performance.	
Total (max. 40)	

Strongly Agree (4), Mostly Agree (3), Partially Agree (2), Totally Disagree (1), Do Not Understand (0)

5. RELIABILITY ANALYSIS

Statement	Score 4, 3, 2, 1, 0
Equipment history is maintained for all key pieces of equipment, showing cause of failure and repair work completed.	
Equipment failures are analyzed to determine root cause and prescribe preventive measures.	
Our failure-prevention efforts are mostly successful. We can usually eliminate the problems we focus on without creating new problems.	
Equipment Mean Time Between Failures (MTBF) and process or mechanical availability are logged/calculated/forecast.	
Value–risk studies have been conducted to optimize maintenance programs.	
All equipment has been classified based on its importance to plant operations and safety. The classification is used to help to determine work order priorities and to direct engineering resources. We work on the most critical equipment's problems first.	
Reliability statistics are maintained even though our employees have a good feel for the best and worst equipment.	
Reliability-centered maintenance or other formal analysis is used to determine the optimal maintenance routines to perform on our equipment.	
Total (max. 32)	

Strongly Agree (4), Mostly Agree (3), Partially Agree (2), Totally Disagree (1), Do Not Understand (0)

6. PERFORMANCE MEASURES/ BENCHMARKING

Statement	Score 4, 3, 2, 1, 0
Labor and material costs are accumulated and reported against key systems and equipment.	
Downtime records including causes are kept on key equipment and systems. These records are periodically analyzed to generate continuous-improvement actions.	
The maintenance department has a set of performance indicators that are routinely measured and tracked to monitor results relative to the maintenance strategy and improvement process.	
All maintenance staff has been trained in or taught the significance of the measures we use. Most of us can read the measures and trends and can determine whether we are improving our overall performance.	
All maintenance trades/areas can see and understand the relationship between their work and results of the department overall. If a particular trade/area is weak they can see it and work to correct it.	
Performance measures are published or posted regularly and kept available/visible for all department staff and trades to see and read.	
Internal and/or or industry norms are used for comparison.	
Maintenance performance of "best in class" organizations has been benchmarked and used to set targets for performance indicators.	

Total (max. 32)

Strongly Agree (4), Mostly Agree (3), Partially Agree (2), Totally Disagree (1), Do Not Understand (0)

7. INFORMATION TECHNOLOGY

Statement	Score 4, 3, 2, 1, 0
A fully functional maintenance management system exists that is linked to the plant financial and material management systems.	
Our maintenance and materials management information is considered a valuable asset and is used regularly. The system is not just a "black hole" for information or a burden to use that produces no benefit.	
Our maintenance management system is easy to use. Most of the maintenance department, especially supervisors and trades, has been trained on it and can and do use it.	
Our planners/schedulers use the maintenance management system to plan jobs and to select and reserve spare parts and materials.	
Parts information is easily accessible and linked to equipment records. Finding parts for specific equipment is easy, and the stock records are usually accurate.	
Scheduling for major shutdowns is done using a project management system that determines critical paths and required levels of resources.	
Condition-based maintenance techniques are supported by automated programs for data analysis and forecasting.	
Expert systems are used in areas where complex diagnostics are required.	
Total (max. 32)	

Strongly Agree (4), Mostly Agree (3), Partially Agree (2), Totally Disagree (1), Do Not Understand (0)

8. PLANNING AND SCHEDULING

Statement	Score 4, 3, 2, 1, 0
A plant equipment register exists that lists all equipment in the plant that requires some form of maintenance or engineering support during its life.	
Over 90% of maintenance work is covered by a standard written work order, standing work order, PM work order, PM checklist, or routine.	
Over 80% of maintenance work (preventive, predictive, and corrective) is formally planned by a planner, supervisor, or other person at least 24 hours or more before being assigned to the trades.	
Nonemergency work requests are screened, estimated, and planned (with tasks, materials, and tools identified and planned) by a dedicated planner.	
Realistic assessments of jobs are used to set standard times for repetitive tasks and to help schedule resources.	
A priority system is in use for all work requests/orders. Priorities are set using predefined criteria, which are not abused to circumvent the system.	
Work for the week is scheduled in consultation with Production and is based on balancing work priorities set by Production with the net capacity of each trade, taking into account emergency work and PM work.	
All shutdowns are scheduled using either critical path or other graphical methods to show jobs, resources, timeframes, and sequences.	
Work backlog (ready to be scheduled) is measured and forecast for each trade and is managed at less than 3 weeks per trade.	
Long-term plans (1–5 years) are used to forecast major shutdowns and maintenance work and to prepare the maintenance budget.	
Total (max. 40)	

Strongly Agree (4), Mostly Agree (3), Partially Agree (2), Totally Disagree (1), Do Not Understand (0)

9. MATERIALS MANAGEMENT

Statement	Score 4, 3, 2, 1, 0
Service levels are measured and are usually high. Stock-outs represent less than 3% of orders placed at the storeroom.	
Parts and materials are readily available for use where and when needed.	
Distributed (satellite) stores are used throughout the plant for commonly used items (e.g., fasteners, fittings, common electrical parts).	
Parts and materials are restocked automatically before the inventory on hand runs out and without prompting by the maintenance crews.	
A central tool crib is used for special tools.	
Inventory is reviewed on a regular basis to delete obsolete or very infrequently used items. An ABC analysis is performed monthly.	
Purchasing/Stores is able to source and acquire rush emergency parts that are not stocked quickly and with sufficient time to avoid plant downtime.	
Average inventory turnovers are greater than 1.5 times.	
Order points and quantities are based on lead time, safety stock, and economic order quantities.	
Inventory is controlled using a computerized system that is fully integrated with the maintenance management/planning system.	
Total (max. 40)	

Strongly Agree (4), Mostly Agree (3), Partially Agree (2), Totally Disagree (1), Do Not Understand (0)

10. MAINTENANCE PROCESS REENGINEERING

Statement	Score 4, 3, 2, 1, 0
Key maintenance processes, e.g., planning and corrective maintenance, have been identified, and "as-is" processes are mapped. Those maps are accurate reflections of the processes that are actually followed.	
Key maintenance processes are redesigned to reduce or eliminate non-value-added activities.	
The CMMS and/or other management systems are used to automate workflow processes.	
Process mapping and redesign have been extended to administration and technical support processes.	
Costs of quality and time for maintenance processes are routinely measured and monitored. Activity costs are known.	
Total (max. 20)	

Strongly Agree (4), Mostly Agree (3), Partially Agree (2), Totally Disagree (1), Do Not Understand (0)

Index